Biocontrol of Arthropods
Affecting Livestock
and Poultry

Biocontrol of Arthropods Affecting Livestock and Poultry

EDITED BY

Donald A. Rutz
and Richard S. Patterson

Routledge
Taylor & Francis Group

NEW YORK AND LONDON

First published 1990 by Westview Press, Inc.

Published 2021 by Routledge
605 Third Avenue, New York, NY 10017
2 Park Square, Milton Park, Abingdon, Oxon OX14 4RN

Routledge is an imprint of the Taylor & Francis Group, an informa business

Copyright © 1990 by Taylor & Francis

Library of Congress Cataloging-in-Publication Data
Biocontrol of arthropods affecting livestock and poultry / edited by
  Donald A. Rutz and Richard S. Patterson.
    p. cm.—(Westview studies in insect biology)
  Includes bibliographical references and index.
  ISBN 0-8133-7850-8
    1. Arthropod pests—Biological control. 2. Veterinary entomology.
  3. Rutz, D. A. 4. Patterson, R. S. I. Series.
  SF810.A3B56 1990
  636.089′4432—dc20
                                                                89-24870
                                                                CIP

ISBN 13: 978-0-3670-1347-9 (hbk)
ISBN 13: 978-0-3671-6334-1 (pbk)

Contents

viii

Acknowledgments

The desktop publishing skills and perseverance of Donna Kowalski have made the preparation and editing of this book an easy task for all, authors as well as editors. We sincerely thank her for her exceptional efforts. We also greatly appreciate the editing assistance of Eric Harrington.

Donald A. Rutz
Richard S. Patterson

1. Status of Biological Control for Livestock Pests

Richard S. Patterson

ABSTRACT

Although over half of the annual farm income is derived from livestock and livestock products, most arthropod pests are controlled by pesticides and very little effort has gone into alternative control methods. Less than 100 scientists work in the area of biological control of livestock pests. In order to be very efficient, biological control agents must be incorporated into a total integrated pest management scheme. A review of what biological control organisms look most promising and the critical needs for the future is given.

KEY WORDS: Parasites, predators, pathogens, control strategies

In the United States, as well as many other countries, over half of the annual farm income is derived from livestock and/or livestock products. It is difficult to assess direct arthropod damage to this commodity because few animals exhibit overt damage or die from arthropod feeding and/or diseases transmitted by them. These pests severely affect the animals, making them very susceptible to other stress factors such as adverse climate, inadequate feed, and any pathogens or internal parasites. Thus, animals weakened by arthropod feeding are the first ones that exhibit a drop in milk production, loss of weight or die if they are stressed severely by any other extrinsic factors. Steelman in 1976 estimated the annual dollar loss to the livestock industry in the United States by insects, mites and ticks to be in excess of four billion dollars (Steelman 1976). Today it is probably two to three times that figure; world wide it is in the tens of billions of dollars. In many developing countries livestock can not even be maintained because of excessive feeding

1

and/or disease transmissions by arthropods. For example, cattle production is very limited on the islands of Zanzibar and Mauritius as well as in many continental African countries because of excessive stable fly (*Stomoxys nigra* Marquat) (Patterson 1989, Zumpt 1973) populations. Cattle and horses often can not be kept in many areas of Africa because of nagana which is spread by tsetse flies (*Glossina* sp.) (Politzar & Cuisance 1982). The same is true of the various livestock diseases which are spread by ticks in various parts of the world. Mosquito populations in the rice growing regions of the world are so dense at times that cattle die or are weakened by the continuous blood feeding (Williams et al. 1985).

Man's health and welfare are also affected by the same arthropod pests which affect livestock, it is even more difficult to put a cash value on man's discomfort and potential disease threats caused by these pests. However, the National Pest Control Association states that commercial pest control is a three billion dollar a year industry in the U.S.A., just to control arthropod pests which invade and attack our homes, ourselves, and our pets. The over counter sales of products to control these same pests is estimated to be another one to two billion dollars annually and growing rapidly (Koehler et al. 1990). Millions of dollars are spent each year on mosquito suppression throughout the world because of their potential damage as disease vectors of humans and livestock. Plus, over a half a million dollars are lost annually, according to the federal government, due to the imported fire ant directly damaging livestock and pastures (Adams 1985).

The demographics of the United States have changed dramatically in the last forty years since World War II. Almost 90% of the voting U.S. public now lives in an urban environment, which they want free of unwanted arthropod pests, yet also desire a clean environment unpolluted from pesticides. The public is also better educated and is quick to question any potential public health nuisance. Present day communication technology has also contributed to this scenario. The public is immediately aware of any public health hazard such as the reports of the spread of lyme disease by its tick vectors (Spielman et al. 1985) or the killing of horses after they had drunk water contaminated with pesticides in 1985 in Jacksonville, Florida (Anon. 1985). Even now there is the very real concern by the public of any wide scale aerial spraying of pesticides for any arthropod control regardless of the crop or the potential economic or public health threat that the pest might have on the state's economy. Therefore, the general public wants an alternative to pesticides, and biological control organisms fit the bill in the public's eye. Unfortunately, it is not that simple. Most biological control organisms must be used in an integrated pest management system which often includes the use of some pesticides.

The management of an arthropod pest which affect man and animals is not a simple process. An entire ecological unit is often treated as a whole just to control a dominant insect pest species. Often many other species of potentially important arthropod pests with unique biological traits are involved. This is further complicated by different livestock management and production schemes of the various animal species. Arthropod control systems, be they biological, chemical or ecological, do not always achieve the same degree of control in different areas of the world because of climatic, cultural and livestock management differences (Axtell 1986). Even in the continental U.S.A. these differences occur for various reasons. Therefore, what works for mosquito, stable fly, or tick control in California does not necessarily achieve the same degree of control of these same pests when used in New York or Florida. The only way to achieve satisfactory control is to develop an arthropod pest management system unique to the particular region that is compatible with current animal production practices.

At present, no single control practice will solve all the livestock arthropod pest problems. In fact, in the past the excessive use of pesticides to achieve this end has often complicated the problem by creating a barren habitat void of all natural parasites, predators, and pathogens as well as having the host become resistant to the pesticide. Thus, a worse problem than before was created because there are no natural suppressing organisms present in the environment. In many cases, a high degree of control can easily be achieved by simply reintroducing parasites, predators and pathogens back into the sterile environment caused by the excessive use of pesticides. Then the habitat must be managed properly in a way that is not conducive to the arthropod pest production (Dietrich 1981). This may be very difficult as most current animal production systems are evolving into larger facilities with dense animal populations often in close proximity to urban centers. Therefore, waste products from the animals, as well as their food, can create serious muscoid fly problems (Hogsette 1981). If this material gets into water, it becomes excellent mosquito breeding habitat (Rutz et al. 1980). When this waste material is handled properly through good manure and/or waste management, then most of the actual or potential fly problem can be eliminated. Since most parasites and predators thrive in favorable habitats, they can further reduce the pest populations below the economic threshold (Ho 1985). Unfortunately habitat modification as used successfully in isolated cases for tick or tsetse fly suppression, and often for flies and mosquitoes, is not always possible because of cost and/or environmental conditions. Thus, other pest control practices must be implemented; unfortunately, little research outside of laboratory studies has gone into developing any type of control other than chemicals. Research has gone into the development of biological control agents for mosquito control and several have been developed commercially. The bacterial toxins of the various strains of *Bacillus thuringiensis* (BT) are the most famous but these are really biological pesticides and not self-perpetuating in most cases. Some of the protozoan diseases and nematodes have shown promise for mosquito control but are not being sold commercially (Lacey & Harper 1986). Various hymenopteran parasites and predaceous mites have been evaluated for muscoid fly control. Some of these parasites have been produced and sold commercially in the U.S.A. and in other countries. The parasites can do a good job if sufficient numbers are released in relation to the number of fly pupae present (Weidhaas & Morgan 1977). This usually takes large numbers, making such releases uneconomical. The best results have been where parasites and predators have been reintroduced into a "sterile" environment, good "waste management" practices have been used, and the indiscriminate use of pesticides has been curtailed.

The state of the art in biological control of arthropod pests of livestock is still in its infancy compared to biological control of plant pests. As stated earlier, it is often difficult to access the damage done by insects, mites and ticks to livestock, but one can easily observe plant damage. Since plants are usually arranged in a large monoculture (corn, wheat, cotton, citrus, etc.), arthropod pest populations can build up to tremendous numbers quickly and biocontrol organisms, especially pathogens, can cause epizootics in the pest population. Even the release of predators, such as mites or beetles, can have very dramatic effects on suppressing pest species of crops since the pests are in such close proximity to one another.

Another factor to consider is that the adult stage of the arthropod pest attacks livestock and man, whereas in the case of most plant pests, the immatures are the major pestiferous stage. It is the immature stages that are most vulnerable to natural suppression factors. In fact over 90-95% of immature stages of most arthropod populations die prior to becoming adults. The adult stage being more mobile is not as vulnerable to biological control organisms such as pathogens, parasites or even predators and survives better. Therefore, in order to raise the efficiency of biological control for suppression of livestock arthropod pests, the emphasis must be made to attack the pest organism in the immature stage. Since a large number of immatures die naturally, a concentrated effort should be made to increase the efficiency of the indigenous biological control organisms, or introduce exotic ones that will compliment the existing beneficial flora and fauna at the breeding sites. Attempts to suppress any adult arthropod pest population with biological control organisms usually is not very practical. If the efficiency of these existing agents can be improved or augmented, the indigenous pest population can easily be maintained within the economic threshold. However, great care must be taken not to upset the natural balance of other beneficial organisms in the environment, otherwise a situation can be created where the biological control agent becomes the nuisance or pest as has happened with many predators. Fire ants can suppress a tick or fly population but they themselves are often a worse menace to livestock than the original pest (Reagan 1986). Since fire ants do not inhabit wooded or bushy areas their efficiency for overall tick or fly control is minimal. We must learn the interrelationship of organisms in the field and this can best be approached through simulated population dynamics models of the pest to be controlled, its potential biological controlling agents, and competitors in a total IPM approach.

The number of entomologists working in biological control of plant pests compared to those working on biological control of livestock pests is very dissimilar. Those working on biological control of livestock pests are a very elite group. There are less than 100 in the entire world and most are in the U.S.A. working on mosquitoes in an augmentation capacity. In the U.S. federal government, only sixteen researchers out of 210 scientists in the field are involved in biological control work on livestock pests. Some of these are only marginally involved because they are taxonomists, geneticists, molecular biologists or scientists using the arthropod parasite or pathogen as models for other systems such as host-parasite interaction, pheromone attractancy, repellency or defense mechanism, etc. It is not much better in the university system; in the U.S.A., less than 8% of the faculty working on biological control of arthropod pests conducts any type of research on livestock pests. Again much of that work is very basic in nature and involves mosquito models to test a theory or supposition. There is almost no effort going into the biological control of ticks or mites, yet these are very important pests of man and livestock and potentially dangerous disease vectors.

In the past two decades, the scientific interest in biological control for the arthropod pests of man and animals has mainly been confined to basic research on pathogens and predators of mosquitoes, competitors (dung beetles), and parasites of muscoid flies. Some effort is now being made in new areas of research, such as pathogens of flies, basic research into strain differences of parasitoids which attack flies, mass releases of parasitoids for fly control, plus the improved virulence of mosquito and fly pathogens. There is some effort in foreign exploration for potential biological control agents for use in fly and mosquito suppression.

For mosquito control, the most intensive research effort with pathogens has been on the spore forming bacteria, *Bacillus thuringiensis* and *B. sphaericus*. Both can be mass produced, formulated and stored and are sold commercially. The bacteria are generally used as microbial insecticides, but *B. sphaericus* can persist for months in some larval habitats. Research on the nature of this persistence is needed, as well as its usefulness. The mode of action of *B. sphaericus* toxin, not yet understood, needs to be elaborated. As basic studies of pathogenic bacteria progress, the microbial insecticides offer great promise for inclusion into IPM systems.

The fungi *Metarhizium, Lagenidium, Entomophthora* and *Coelomomyces* are all pathogenic to mosquitoes and their differing characteristics suggest possible bio-control roles in specific mosquito and biting gnat habitats. A better understanding of the pathogenicity and mode of entry of the fungal infective units is needed. Resistant spores capable of long term storage and of systems for production and dissemination of more fragile infectious forms are required. Efficacy of fungi is variable and activity can be lost or enhanced by fermentation methods. The biological basis for this variability requires investigation.

Among the protozoa, the microsporidia are probably the most common pathogens in mosquitoes and black flies, and continued scientific effort is required in order to determine their mode of infection; this needs to be fully elucidated before the bio-control potential of these organisms can be examined. Viruses, although common and often virulent in nature, have proven to be of low infectivity and are difficult to mass produce; as such, they represent a very high risk area of research, but should not be ignored (Roberts & Castillo 1980).

Extensive research has gone into the development of nematodes for mosquito control with very promising results. The techniques are on the verge of being commercially accepted. Nematodes work very well on clean water mosquitoes such as many of the Anopheline and Aedes species (Petersen 1975). This is also true of the research on *Toxorhynchites* spp., the predatory mosquito, being used to suppress *Aedes aegypti* populations, but more research needs to be done so that these two technologies can be fully implemented into mosquito control programs (Focks et al. 1980).

The critical research needed in biological control of muscoid flies which affect man and animals is to evaluate the efficiency of the various bio-control agents and their interrelationships with one another and the various host populations, especially when multiple hosts are present. This is not well understood. Although the hymenopterous parasitoids of the various genera and species are cosmopolitan in distribution, their efficiency in attacking the hosts varies depending on geographic region, climate, season, habitat, host density and host distribution. For example, *Nasonia vitripennis* appears to be a very efficient parasite in the New York area in attacking house fly pupae, but exhibits a very poor efficiency in similar facilities in Florida. The same is true of *Muscidifurax* spp. It is the parasite of choice in California, but is relatively inefficient in the same type of habitat in Florida where *Spalangia* spp works best against muscoid flies. All these species are present in each area, one or two outcompeting the others due to some subtle differences. An evaluation of strain differences of the same species collected from different locations in the U.S. and abroad must be done in the laboratory and the field. Using an electrophoretic analysis of isoenzymes of the parasitoids will help us to: (1) resolve taxonomic relationships of previously indistinguishable groups, (2) provide evidence for novel genes in exotic material, and (3) provide a means (i.e., marker alleles) to evaluate establishment and post-colonization performance of exotic parasites and predators.

The biggest obstacle is the evaluation of the efficiency of these parasitoids under various regional, climatic and habitat type. The efficiency of each strain and species must be evaluated alone and together in various locales in the U.S.A. and abroad to determine which ones are most efficient under each condition.

8

Much more research effort must go into understanding the role predators play in suppressing certain pest populations; almost no research effort is going into this area, yet in many incidences predators are main biological suppressing agents of pest populations. The same is true of competitors; their role must be determined. For years it was assumed that dung beetles would reduce fly populations because they broke up and buried the dung; studies now show that as a fly controlling agent, the dung beetles are not very efficient. Some competitors appear to have a significant effect on pest populations and may reduce them to below an economic threshold level. Insufficient studies have been conducted to prove the point on most species. In some cases the competitors may even act as an intermediate host for various pathogens of the target pest.

A few basic or preliminary studies have been conducted on the many aspects of biological control of arthropods which affect livestock, but few long term, in depth studies have been conducted. Very little has been done to develop techniques of habitat modification that will be beneficial to the production of parasites, predators and pathogens of the pest species. In the next decades, new biological control techniques must be developed which are economically feasible to suppress livestock arthropod pests because few new pesticides will be available in the future for use on or near livestock.

REFERENCES CITED

Adams, C. T. 1986. Agricultural and medical impact of the imported fire ants (48-57). In Fire Ants and Leaf-Cutting Ants Biology and Management. Ed. C. S. Lofgren and R. K. Vander Meer. Westview press, Boulder. 435.

Anon. 1985. "Horses die of pesticide poisoning." Jacksonville Times. August 27,1985.

Axtell, R. C. 1986. Status and potential of biological control agents in livestock and poultry pest management systems. In Biological Control of Muscoid Flies. Ed. R. S. Patterson and D. A. Rutz. Entomol. Soc. Misc. Publ. 61:1-9.

Dietrick, E. J. 1981. Commercial production and use of predators and parasites for fly control programs. pp. 192-200. Status of Biological Control of Filth Flies. USDA, SEA, Publ. A 106.2F 64. 212p.

Ho, C. C. 1985. Mass production of the predaceous mite, *Macrocheles muscaedomestice* (Scopoli) (Acarina: Macrochelidae), and its potential use as a biological control agent of house fly, *Musca domestica* L. (Diptera: Muscidae). Ph.D. Diss. Univ. of Florida. 185p.

Hogsette, J. A. 1981. Fly control by compositing manure at a south Florida equine facility. In Status of Biological Control of Filth Flies. USDA, SEA, Publ. A 106.2F 64. 105p.

Koehler, P. G., R. S. Patterson, & R. J. Brenner. 1990. Cockroaches. In Malis, A. Handbook of Pest Control. Ed. K. Story. Franzak & Foster, Cleveland, Ohio. pp. (in press).

Lacey, L. A., & J. D. Harper. 1986. Microbial control and integrated pest management. J. Entomol. Sci. 21:206-213.

Patterson, R. S. 1989. Biology and Ecology of *Stomoxys nigra* and *S. calcitians* on Zanzibar, Tanzania. pp. 2-11. In Current Status of Stable Fly (Diptera: Muscidae) Research Miscellaneous Publication 74. Ed. J. J. Petersen & G. L. Greene. Entomol. Soc. of America, College Park, MD.

Petersen, J. J. 1975. Status of nematodes as mosquito control agents in North America. Prox. Alberta Mosq. Abat. Symp. pp. 181-190.

Politzar, H. & D. Cuisance. SIT in the control and eradication of *Glossina palpalis gambiensis*. "Sterile Insect Technique and Radiation in Insect Control." (Proc. Symp. Neuherberg 1981) IAEA Vienna (1982) 101.

Reagan, T. E. 1986. Beneficial aspects of the Imported Fire Ant: A Field Ecology Approach. In Fire Ants and Leaf-Cutting Ants Biology and Management. Ed. S. C. Lofgren and R. K. Vander Meer. Westview Studies in Insect Biology.

Roberts, D. W. & J. M. Castillo, eds. 1980. Bibliography on pathogens of medically important arthropods 1980. Suppl. No. 1 to Vol. 58. Bull. Wld. Hlth. Org. 205 pp.

Rutz, D. A., R. C. Axtell, & T. E. Edwards. 1980. Effects of organic pollution levels on aquatic insect abundance in field pilot-scale anaerobic animal waste lagoons. Mosquito News 40:402-409.

Spielman, A. M., M. L. Wilson, J. F. Levine, & J. Piesman. 1985. Ecology of *Ixodes dammini* borne human babesiosis and lyme disease. Annu. Rev. Entomol. 30:439-460.

Steelman, C. D. 1976. Effects of external and internal arthropod parasites in domestic livestock productions. Ann. Rev. Entomol. 21:155-178.

Weidhass, D. E. & P. B. Morgan. 1977. Augmentation of natural enemies for control of insect pests of man and animals in the U.S. Bio. Con. by Augmentation of Natural Enemies. 14:417-428.

Williams, R. E., R. D. Hall, A. B. Broce, & P. J. Scholl. 1985. Livestock Entomology. Wiley. Interscience, New York, N.Y. p. 335.

Zumpt, F. 1973. The stomoxyine biting flies of the world Diptera: Muscidae. Gustor Fascher Vulag, Stuttgart, Federal Republic of Germany. p.175.

2. Host-Parasite Relationship of *Dirhinus pachycerus* Masi (Hymenoptera: Chalcididae), with Particular Reference to Its House Fly Control Potential

M. Geetha Bai

ABSTRACT

Studies on various aspects of the biology of *Dirhinus pachycerus,* a pupal parasitoid of house flies in the Indian sub-continent are presented. The aspects dealt with include mating and insemination, oviposition and host-feeding, daily rate of oviposition, fecundity, sex-ratio, ovigenic cycle, adult nutrition and longevity. Possibilities of utilizing this parasitoid in integrated house fly control programs are discussed.

INTRODUCTION

Dirhinus spp. are of great economic importance from the point of view of biological control of several insect pests. They are generally pupal parasitoids of some members of the Families Calliphoridae, Glossinidae, Muscidae, Sarcophagidae and Trypetidae. *D. pachycerus* Masi is a solitary external parasitoid of some calliphorid, muscoid and sarcophagid flies. This species was described from a few specimens collected at Calcutta in India by Masi in 1921, but its host was unknown. It was later reared from puparia of *Sarcophaga dux* var. *tuberosa* Pand. in the same area (Roy et al., 1940). Bionomics of *D. pachycerus* in India has been studied (Geetha Bai & Sankaran, 1982). Some aspects of studies on the biology of *D. pachycerus* are presented.

MATERIALS AND METHODS

<u>Mating and Insemination.</u> Parasitized puparia of *M. domestica*, with the parasitoids in late pupal stage, were cut open at

11

the anterior end and isolated in closed glass vials (3.0 X 5 cm.) and mating behavior was observed after emergence. Freshly emerged virgin females were provided to unmated males every day until their death.

Oviposition and Host Feeding. Five-day old gravid females, fed on 50% honey solution in water, were provided with ten, twenty-four to forty-eight hour old puparia, of *M. domestica* pasted on a card and kept under observation until all hosts were attacked. The experiment was replicated ten times. To study the frequency of host-feeding, five females fed on honey and five not fed were given ten, twenty-four and forty-eight hour old host puparia and kept under observation for four hours, daily from the first day until the death of the parasitoids. All the hosts were dissected to confirm if they were parasitized.

Daily Rate of Oviposition, Fecundity, Longevity and Sex-Ratio. Ten freshly emerged female parasitoids were allowed to mate with freshly emerged unmated males. Each female was fed on 50% honey solution and provided with twenty, twenty-four to forty-eight hour old puparia of *M. domestica*, every 24 hours from the first day of its life until death. These puparia were kept under observation for emergence of flies or adult parasitoids. The number and sex-ratio of progeny of all ovipositing females were recorded. Fly puparia from which neither flies nor parasitoids emerged after three months were dissected to examine their contents.

Ovigenic Cycle. The ovigenic cycle of mated females of two different size groups, viz., 4.3-4.6 mm long, and 2.8-3.1 mm long, one batch fed on 50% honey solution and deprived of hosts and the other starved and deprived of hosts, was studied. Ovaries were dissected in distilled water mixed with an equal part of Bouin's fluid to fix the material and then mounted in glycerine. Acetocarmine imparts a red color to developing and resorbing eggs while mature eggs do not take up the stain. Stained oocytes less than about 0.3 mm long were grouped as (a) early stage of maturation; stained oocytes more than 0.3 mm as (b) late stage of maturation; unstained eggs with the typical shape (c) as mature eggs, and stained eggs at the base of the ovariole with irregular outline as (d) resorbing eggs.

Adult Nutrition and Longevity. The effect of nutrition on longevity of adult parasitoids was studied by feeding twenty mated females and males on the following diets:

50% honey solution, mucilage obtained from *Hibiscus rosasinensis* (Malvaceae) nectar from *Leucas aspera* (Labiatae), raisins or sucrose solution. After mating the parasitoids were isolated in glass vials (5.5 X 1.0) and fed every day on the various diets.

RESULTS AND DISCUSSION

Mating and Insemination. Because *D. pachycerus* is arrhenotokous, only those females which produced female progeny were obviously successfully inseminated, while those which produced only males were not. Both males and females mate soon after emergence. Females are uninuptial, but male parasitoids mate with a number of females in succession. Males mated the maximum number of times on the first day of their life, the maximum and minimum being thirty-six and seventeen, respectively (Table 1). A gradual decrease in mating frequency was noticed on the second and third day followed by a steep decline later. The total number of matings during the whole lifetime ranged from thirty-six to 107, with an average of 72.8. The total number of females that were inseminated by a male ranged from thirty-one to eighty-four, with an average of 62.2. One of the males mated thirty-six times on the first day and died on the third day. All the mated females were not inseminated. Of the 364 females used in these studies, only 62% were found inseminated. The longevity of males ranged from three to eighteen days, with an average of 9.8 days.

Oviposition and Host Feeding. A gravid female in search of host puparia walks at random with her antennae touching all objects she comes across. When her antennae touch a fly puparium, the parasitoid immediately mounts it and starts examining it with her antennae (Fig. 1a). If she finds the host unsuitable she leaves it, and if found suitable, she taps it with the tip of her abdomen (Fig. 1b). When she finds a suitable site for oviposition, she presses it with her stylets, unsheaths her ovipositor and starts drilling a hole. If she is not successful, she moves to another spot and repeats the process. When the puparial covering has been pierced, she rotates the ovipositor, while pushing it down and partially pulls it out and withdraws the stylets (Fig. 1c). If she finds a suitable site on the host, she lays a single egg on the pupa and withdraws the ovipositor. Generally the anterior end of the fly puparium is preferred for oviposition. *D. pachycerus* females normally drill a single hole, but if that site is not suitable she drills several holes. Of the 100 puparia tested, a single hole was drilled in eighty-three puparia, two holes in thirteen, three holes in two hosts, four and six holes in one each. Seventy-nine hosts drilled once were found

Table 1. Insemination by *Dirhinus pachycerus* males.

Sl. No. of males	Day 1	2	3	4	5	6	7	8	9	10	11	12	13	14	Total no. of matings	No. of females inseminated	Longevity of copulating males
1	30	25	19	4	8	1	3	–	–	–	–	–	–	–	90	79	12
2	20	18	27	12	9	1	4	–	–	–	–	–	–	–	91	84	10
3	17	15	16	8	11	9	4	–	9	–	9	8	–	1	107	82	18
4	36	–	–	–	–	–	–	–	–	–	–	–	–	–	36	31	3
5	20	11	9	–	–	–	–	–	–	–	–	–	–	–	40	35	6
Total	123	69	71	24	28	11	11	–	9	–	9	8	–	1	364	311	
Av.	24.6	13.8	14.2	4.8	5.6	2.2	2.2	–	1.8	–	1.8	1.6	–	0.2	72.8	62.2	9.8

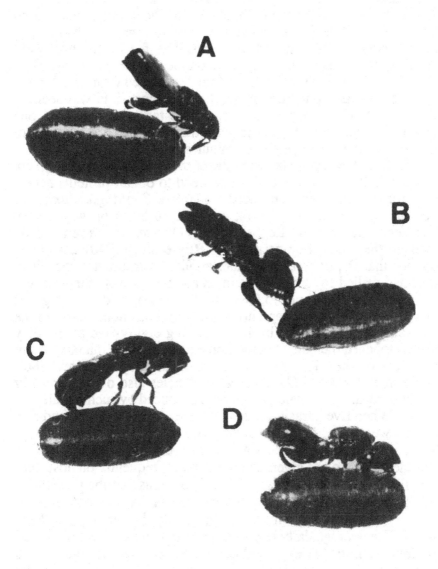

Figure 1. Oviposition and host feeding in *Dirhinus pachycerus*. a. Drumming; b. Tapping; c. Ovipositor pierced through the puparial wall; d. Host feeding.

parasitized while no eggs were found in the remaining four. In these four cases the parasitoids withdrew their ovipositor quickly and left the host probably because the host pupa was found unsuitable. All the hosts drilled two, three or four times had an egg in the spots drilled last and the parasitoids fed on the host fluid which appeared as a whitish globule on the surface of the puparium (Fig. 1d). Sometimes ovipositing females applied their mouth parts to the oviposition puncture even though no host fluid was seen on the surface of the puparium. The time taken to complete one oviposition varied from three to sixty-three minutes, the average for fifty ovipositions being eighteen minutes.

Host-feeding is characteristic of many parasitic Hymenoptera (Flanders, 1935). The females host-feed to obtain proteins necessary for optimum egg production. Unlike *Spalangia cameroni* Perkins, where females do not oviposit on hosts on which they host-feed (Gerling and Legner, 1968), *D. pachycerus* females do feed on the hosts they parasitize. Roy et al., (1940) are of the opinion that *D. pachycerus* females applied their mouth parts where they had drilled for oviposition in an attempt to close the puncture with their saliva and not to feed. In the present study, among 100 ovipositions observed, parasitoids applied their mouth parts to the hole only fifteen times. In the remaining eighty-five cases, they merely touched the hole with their antennae for a few minutes before they left or they left the host as soon as they withdrew their ovipositor. Unmated *D. pachycerus* females also oviposit readily and host-feed. Only male progeny are produced by such females.

When five parasitoids fed on honey throughout life and five not fed were provided with puparia and their oviposition behavior was observed for four hours a day from the first day of their life until death, those not given honey fed more frequently on host fluid thán those given honey (Table 2). It is obvious that the parasitoids not fed on honey depend more on host exudate for their nutrition.

Daily Rate of Oviposition, Fecundity, Longevity, and Sex-Ratio. The average daily rate of oviposition and the maximum and minimum number of eggs laid daily are presented in Fig. 2. The average oviposition rate increased from the first day, reached its peak on the seventh and eighth days. The maximum number of eggs laid being seventeen on the seventh and eighth days. After this period, oviposition rate decreased until the thirteenth day and again increased. The second peak was observed on the sixteenth day. Oviposition rate gradually decreased after this period, except for small peaks and finally ceased on the thirty-sixth day. Details of fecundity, longevity, and sex-ratio are presented in Table 3. The

Table 2. Frequency of host-feeding in _Dirhinus pachycerus_.

Sl. No. of parasitoid	Honey fed; hosts offered				No honey; hosts offered			
	No. of hosts attacked	No. of hosts parasitized	No. of times host fed	Longevity (days)	No. of hosts attacked	No. of hosts parasitized	No. of times host fed	Longevity (days)
1	18	17	4	16	30	30	8	30
2	21	21	5	25	29	29	7	34
3	41	41	5	39	36	36	7	36
4	22	22	2	32	27	27	5	31
5	39	39	3	45	31	31	8	30
Total	141	140	19	157	153	153	35	161
Average	28.2	28	3.8	31.4	30.6	30.6	7	32.2

Table 3. Longevity, fecundity and sex-ratio of *Dirhinus pachycerus*.

Repli-cate No.	Longevity of ovipositing females (days)	No. of fly puparia parasitised	No. of adult parasitoids emerged			%Sex-ratio		Total % mortality of developing parasitoids	Pre-Ovi-position	Post-Ovi-position
			male	female	Total	male	female			
1	52	154	46	72	118	39	61	23.4	3	24
2	44	128	10	87	97	10	90	24.1	2	5
3	39	128	16	93	109	15	85	14.8	2	18
4	31	116	19	67	86	22	78	25.8	1	9
5	40	159	70	58	128	55	45	19.8	1	12
6	33	120	28	68	96	31	69	25.0	2	9
7	41	141	29	78	107	27	73	24.8	3	10
8	43	132	19	67	86	22	78	34.8	3	8
9	43	126	11	85	96	11	89	28.8	2	11
10	51	150	32	86	118	27	73	21.3	3	14
Average	41.7	135.4	28.0	76.1	104.1	25.9	74.1	24.3	2.2	12.0
Range	31	116	10	58	86	10	45	14.8	1	5
	-52	-159	-70	-93	-128	-55	-90	-34.8	-3	-24

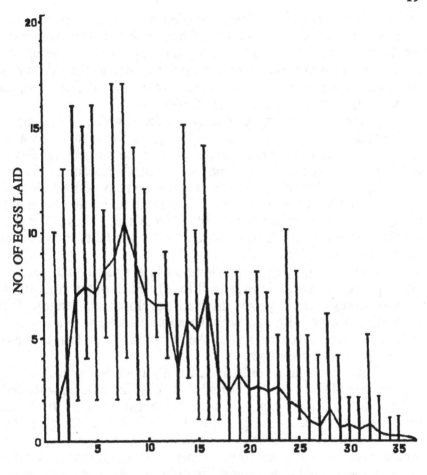

Figure 2. Average daily rate of oviposition in *Dirhinus pachycerus* (Vertical lines indicate the maximum and minimum number eggs laid).

total number of eggs laid during the lifetime ranged from 116-159. The maximum observed longevity of ovipositing females was fifty-two days and the minimum was thirty-one days, with an average of 41.7. The pre-oviposition period ranged from one to three days and averaged 2.2 days. and the post-oviposition period from five to twenty-four days, with an average of twelve days.

The progeny produced during the first few days consisted of both females and males, the former being predominant. Towards the end of a parental females' life, the sex of the progeny produced by her was mostly male. Only males were produced during the last few days. Totally more female progeny were produced in nine replicates, while in one case only 45% were females, the remaining being males. The maximum mortality of immature stages of the parasitoid was 34.8% and the minimum 24.8%. Mortality in the pupal stage was 10.8-25.9%.

Ovigenic Cycle. *D. pachycerus* is a synovigenic species with a pair of polytrophic ovaries (Fig. 3). Each ovary consists of three ovarioles with oocytes in different developmental stages. The developing oocytes alternate with the nurse cells. The ovarioles on each side lead into the oviduct and the two oviducts unite at the base to form the common oviduct.

The larger parasitoids lived for a longer duration than the smaller ones (Tables 4 & 5). Females in both size groups, when fed on 50% honey, lived longer than starved ones. Most of them had larger number of developing and mature ova than those given no food. The number of developing ova and mature eggs in smaller females was generally less than in larger ones. Mature ova were found on the second day of their life both in fed and starved larger females. In smaller females, mature ova were found on the second day in fed and on the third day in starved ones. Among larger females resorption of eggs in fed and starved ones started when they were ten days and six days old, respectively. The corresponding figures for smaller females were seven and five days, suggesting that resorption of eggs starts earlier in smaller females than in larger ones. As the females advanced in age, the number of mature eggs decreased and the number of resorbing eggs increased. The number of ova is related to size and nutritional state of female *D. pachycerus*. Since the size of parasitoids is directly proportional to the size of their hosts, it is desirable to use larger hosts to obtain parasitoids with higher fecundity and longevity. Edwards (1954) and Wylie (1966) also reported a reduction in longevity of smaller females of *Nasonia vitripennis* (Walker).

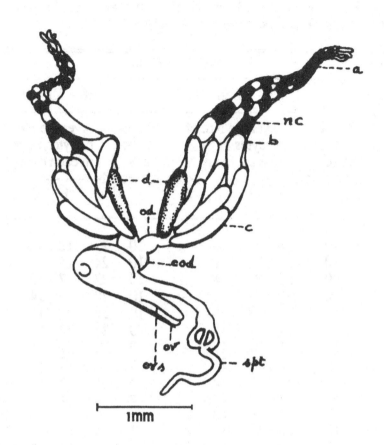

Figure 3. Female reproductive system of a starved eight-day old *Dirhinus pachycerus.* cod = common oviduct; nc = nurse cells; od = oviduct; or = ovipositor; ors = ovipositor sheath; spt = spermatheca.

Table 4. Condition of ovaries of mated and honey-fed *Dirhinus pachycerus* females deprived of hosts.

Age of para-sitoids (days)	Size of parasites									
	4.3 - 4.6 mm					2.8 - 3.1 mm				
	No. of parasi-toids dissected	Average No. of oocytes/eggs in each ovary				No. of parasi-toids dissected	Average No. of oocytes/eggs in each ovary			
		a	b	c	d		a	b	c	d
0	10	38	14	-	-	10	27	18	-	-
1	10	31	21	6	-	10	24	15	-	-
2	10	33	24	8	-	10	22	17	6	-
3	10	29	27	11	-	10	20	16	14	-
4	10	37	28	16	-	10	21	22	14	-
5	10	24	24	18	-	10	24	22	15	-
6	10	25	22	17	-	10	22	21	16	-
7	10	19	23	21	-	5	18	19	10	6
8	10	18	20	23	-	5	17	18	12	8
9	10	17	18	22	-	5	12	17	9	11
10	10	18	14	20	2	3	14	14	6	14
11	10	19	17	17	6	2	12	16	7	12
12	5	17	18	15	5	2	10	15	5	7
13	5	16	16	16	4					
14	5	13	15	8	7					
15	5	15	17	10	9					
16	5	14	15	9	11					
17	5	14	12	8	10					
18	5	12	12	11	14					
19	5	13	10	10	12					
20	5	13	14	7	16					
21	5	10	18	5	15					
22	3	14	15	-	14					
23	2	15	16	-	16					
24	2	9	17	-	16					
25	1	8	19	-	18					

Table 5. **Condition of ovaries of mated and starved** *Dirhinus pachycerus* **females deprived of hosts.**

Age of para-sitoids (days)	Size of parasites									
	4.3 - 4.6 mm					2.8 - 3.1 mm				
	No. of parasitoids dissected	Average No. of oocytes/eggs in each ovary				No. of parasitoids dissected	Average No. of oocytes/eggs in each ovary			
		a	b	c	d		a	b	c	d
0	10	34	22	-	-	10	21	14	-	-
1	10	29	20	22	-	10	17	15	-	-
2	10	31	17	4	-	10	16	14	-	-
3	10	28	22	7	-	10	16	18	3	-
4	10	24	19	14	-	5	15	16	6	-
5	10	26	14	11	-	5	15	14	4	3
6	10	21	16	19	2	5	16	10	8	6
7	10	18	18	16	3	5	12	9	9	4
8	10	19	15	18	2	5	10	7	5	7
9	5	14	20	16	4	2	12	8	6	9
10	5	16	19	15	6					
11	5	14	16	14	4					
112	5	10	15	16	7					
13	5	11	16	12	9					

Adult Nutrition and Longevity. Adult nutrition has a profound influence on survival, longevity, and fecundity of insects. In laboratory colonies adult parasitoids are generally fed on dilute honey, sucrose solution, glucose or other carbohydrates. In nature many insect parasitoids feed on dew, nectar or mucilaginous secretions of plants.

Mated females fed on raisins lived for the longest duration followed by those fed on nectar, 50% honey solution, mucilage, and sucrose solution, in the given order (Fig. 4). Mated males fed on nectar lived for the longest duration followed by those fed on

24

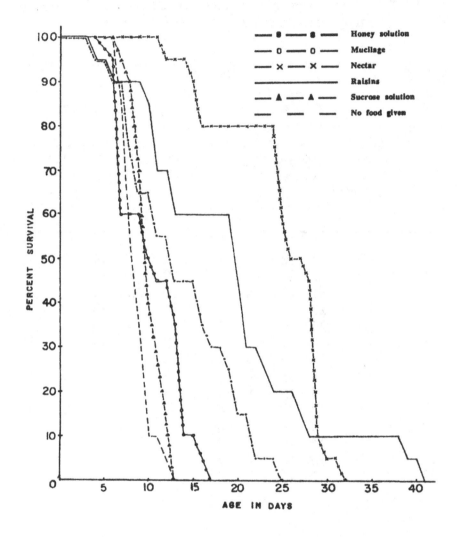

Figure 4. Longevity of mated females fed on different diets.

raisins, sucrose solution, and mucilage in this order (Fig. 5). Starved as well as fed mated females lived longer than starved or fed mated males. When twenty unmated males were fed on 50% honey solution, they lived for the longest duration compared with mated males fed on various diets. It is thus observed that adult nutrition has a profound effect on longevity of *D. pachycerus*. Starved adult parasitoids lived for the shortest duration compared to those given some kind of food.

CONCLUSION

Detailed studies carried out on the biology of *D. pachycerus*, a solitary parasitoid of *M. domestica* recorded so far only from India and Pakistan, have added to our knowledge of the biology of one of the important natural enemies of house flies, which will help in its evaluation for use in fly control programs. Being a synovigenic species, it is preferable to pro-ovigenic ones for field releases (Flanders, 1950). Ovipositing females kill their hosts by stinging them prior to deposition of eggs. Even though the developing parasitoids sometimes die during development, every attack by the ovipositing females results in the death of the developing fly pupa. Female parasitoids derive nutrition from feeding on the host blood through the hole drilled for oviposition.

D. pachycerus has a very limited geographical distribution and therefore possibilities of introduction of this parasitoid in other parts of the world, especially the USA, where natural enemies are exploited in fly control programs, may be explored. This parasitoid would add to the already existing natural enemy complex and thus help the effect of biocontrol agents in controlling house flies. *D. pachycerus* is amenable to mass production in the laboratory. This parasitoid also accepts frozen house fly puparia and their progeny successfully complete their development in such hosts (Geetha Bai, 1980). Therefore, fly puparia can be stored by freezing them and can be used at appropriate times to mass produce parasitoids for augmentative releases.

ACKNOWLEDGMENTS

The author would like to express her deep sense of gratitude to Dr. T. Sankaran for help and guidance during the project and for critically going through the manuscript. Financial assistance of the

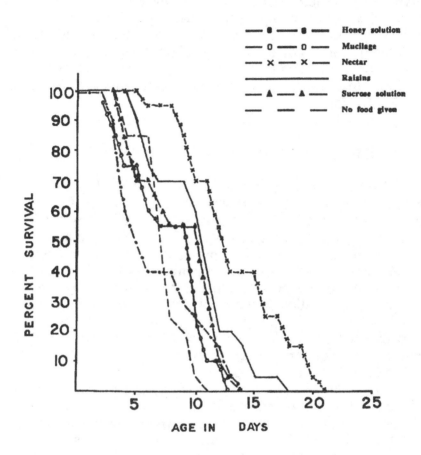

Figure 5. Longevity of mated males fed on different diets.

Indian Council of Medical Research, New Delhi during the duration of this study is gratefully acknowledged.

REFERENCES CITED

Edwards, E. L. 1954. The effect of diet on egg maturation and resorption in *Mormoniella vitripennis* (Hymenoptera: Pteromalidae). Quart. J. Microscop. Sci., 95:459-468.

Flanders, S. E. 1935. An apparent correlation between the feeding habits of certain pteromalids and the conditions of their ovarian follicles (Pteromalidae: Hymenoptera). Ann. Ent. Soc. Amer., 28:438-444.

_____. 1950. Regulation of ovulation and egg disposal in the parasitic Hymenoptera. Can. Ent., 82:134-140.

Geetha Bai, M. 1980. Studies on some natural enemies of House fly and certain other Muscoid Flies. Ph.D. Dissertation, Univ. of Mysore, India.

Geetha Bai, M. & Sankaran, T. 1982. Seasonal occurrence of *Musca domestica* L. (Dipt.: Muscidae) and other flies in relation to their pupal parasites in animal manure in and around Bangalore. Proc. Symp. Ecol. Animal Popul. Zool. Surv. India, 3:133-141.

Gerling, D. & Legner, E. F. 1968. Developmental history and reproduction of *Spalangia cameroni*, parasite of synanthropic flies, Ann. Ent. Soc. Amer., 61:1436-1443.

Roy, D. N. & Siddons, L. B. 1939. A list of Hymenoptera of super family Chalcidoidae, parasites of Calyptrate Muscoidae. Rec. Indian Mus. (Calcutta), 41:223-224.

Roy, D. N.; Siddons, L. B. & Mukherjee, S. P. 1940. The Bionomics of *Dirhinus pachycerus* Masi (Hymenoptera, Chalcidoidea), a pupal parasite of Muscoid flies. Indian J. Ent. 2:225-240.

Wylie, H. G. 1966. Some effects of female parasite size on reproduction of *Nasonia vitripennis* (Walk.) (Hymenoptera: Pteromalidae). Can. Ent., 98:196-198.

3. Biological Control of Filth Flies in Confined Cattle Feedlots Using Pteromalid Parasites

Gerald L. Greene

The largest concentration of confined cattle in feedlots is in the high plains region of the United States. The states of Colorado, Kansas, Iowa, Nebraska, Oklahoma, and Texas produce over half of the fat cattle in the United States. The region from western Texas north to western Nebraska, with Garden City, Kansas, in the center, constitutes the major cattle feeding area, with approximately 20 million head produced yearly. The availability of grain produced with irrigation and relatively dry feedlot conditions throughout the year make this area well adapted to efficient cattle feeding in large feedlots. Those lots feed thousands of cattle in a modern confinement setting. Slaughter plants have been built in the feeding area, with Kansas being the leading state in number of cattle slaughtered, almost seven million per year.

With the major concentration of cattle in the high plains, it is easy to see why large amounts of animal waste accumulate and large populations of flies develop. Cattle in confinement suffer considerable irritation, when the stable flies, *Stomoxys calcitrans* (L.), attempt to take their blood meals. Cattle are irritated to the point of reducing their feed intake, resulting in reduced weight gain. Using figures of Campbell et al. (1977), the loss from stable fly feeding is over $100 million per year in the high plains cattle feedlots. The additional problem of flies (mainly house flies, *Musca domestica* L.) bothering the people working at the feedlots or living close by, mandates that fly control must be improved. Without improved fly control, cattle feeding efficiency continues to be reduced, and feedlots close to cities or population centers may be forced to close.

Biological control of filth flies in cattle feedlots has been attempted for several years, using commercially produced parasitic wasps. However, there has not been research documentation of actual fly reduction or dramatically increased parasitism of fly pupae, following the release of commercial parasites at cattle feedlots in the high plains area. The benefits claimed by commercial producers of these wasps have more often resulted from better sanitation or reduced fly numbers because of drier, less favorable conditions for fly development. The success of Legner & Brydon (1966), Morgan & Patterson (1977), Axtell (1981) in poultry environments created considerable interest in fly control with parasites. Unfortunately, this technology was transferred to cattle feedlots without adaptation to the different environmental conditions, or geographical region. Morgan (1980) reported 84% fly reduction at a beef farm in Florida by releasing large numbers of *Spalangia endius* Walker. Entrepreneurs assumed that similar results would be obtained if *S. endius* were released in high plains feedlots. However, the drier conditions and temporal fly populations in feedlots of this area are quite different from those of poultry houses or Florida beef farms.

With the promotion for sales of parasites, purchases exceeded one million dollars per year in the early 1980's in the high plains. The interest in using parasites to reduce fly populations grew to the point that questions were being asked about which parasite species to use and the level of control that could be expected.

Stage & Petersen (1981) and Petersen et al. (1983) released *S. endius, Nasonia vitripennis* (Walker), and *Muscidifurax raptor* Girault & Sanders in Nebraska and did not record significant fly reductions or elevated parasitism levels compared with those in feedlots with only natural parasitism. In 1982, I experienced similar results with *S. endius* releases in a western Kansas cattle feedlot. In several subsequent release studies, I have retrieved less than 1% *S. endius* from fly puparia collected from release lots. Therefore, the evidence is substantial that *S. endius* is not adapted to the environment in which the majority of American beef cattle are fattened. That is in contrast to reports by Legner et al. (1967) and Legner & Olton (1968, 1970, 1971), who found *S. endius* in many animal waste areas sampled in the eastern and western hemisphere. They reported *S. endius* from both house fly and stable fly puparia. Petersen & Meyer (1983) found that other pteromalid species, *Spalangia cameroni* Perkins, *S. nigroaenea* Curtis, *S. nigra* Latrielle, and *Muscidifurax* spp., were much more common than

S. endius in field collected fly puparia in Nebraska. I have observed *M. raptor*, *M. zaraptor*, Kogan & Legner, *S. cameroni*, *S. nigroaenea*, and *Urolepis rufipes* (Ashmead) to be more common than *S. endius* in Kansas feedlots. We may have substantiated the suggestion by Rutz & Axtell (1979) that unless experimental evidence exists to show that a parasite species is effective in an area of introduction, the shipment of insectary-grown parasites from one climatic region to another is a questionable procedure. None of the stock colonies of parasites released in Kansas and Nebraska were collected from those states.

Some basic questions must be answered before parasite releases can be expected to achieve fly control of high plains cattle feedlots. These include:

1. Which parasite species occur naturally in the dry cattle feedlot environment in the high plains?
2. What impact do natural parasites exert on fly populations?
3. What are the seasonal population dynamics of the natural parasites?
4. Are the methods and rates of release conducive to fly control?
5. Which parasite species are appropriate for mass releases?
6. Would parasite releases provide economical fly control?

In an attempt to answer these questions, I have sampled naturally-occurring fly pupae from cattle feedlots in western Kansas. From 1982 to 1986, twenty-five feedlots were sampled, including six of the same feedlots each year. Those feedlots held from 100 to 75,000 cattle. Samples of approximately 100 fly puparia were collected weekly from May to October from each feedlot and brought to the laboratory. Fly parasite species emergence was then recorded. The sampled area has an average of 45 cm moisture per year, with over half of it received during May-July, a period when stable fly populations are high (Greene, 1989).

Parasites in the Cattle Feedlot

The Pteromalidae recovered were *Muscidifurax* and *Spalangia* in a nearly fifty-fifty ratio, constituting over 95% of all the parasites recovered from nearly 70,000 puparia collected. There were two dominant species; *S. nigroaenea* made up 75% of the

Spalangia, and *M. zaraptor* 79% of the *Muscidifurax*. *S. cameroni* and *M. raptor* constituted the majority of the remaining specimens. This species composition is similar to that reported from Nebraska cattle feedlots by Petersen & Meyer (1983) where *Muscidifurax* spp. made up 56.0% of the house fly parasites and only 25.0% of the stable fly parasites. The stable fly parasites were 43.6% *S. nigroaenea*. The presence of *S. nigroaenea* in Kansas at about 900 m elevation is in contrast to results of Legner & Olton (1971), who found this species below 600 m elevation. Legner (1977) found *Spalangia* parasitized host pupae at greater depths than *Muscidifurax*. This may explain why *Spalangia* is a more abundant parasite of stable fly pupae, which usually occur deeper in the residue than do house fly pupae.

In choosing a biological control agent, we must look at the relative abundance of the parasite species occurring on the major fly species. *M. zaraptor* appears to be predominantly a parasite of house fly pupae (92%) in Kansas and may not be effective for stable fly control. *S. nigroaenea* appears to be the best native parasite for stable flies, because the number of *S. nigroaenea* recovered from stable fly puparia were greater than from house fly puparia in Kansas and Nebraska cattle feedlots (Meyer & Petersen 1982) when field collections contained both house fly and stable fly pupae. Similar observations were made in Florida (Greene et al. 1989).

Population Levels

Twenty-five different feedlots averaged 37.5% parasitism of the puparia producing emerged parasites or eclosed flies (percentage calculated by dividing the number of emerged parasites by the total of the emerged parasites plus eclosed flies) and 17% parasitism of the total number of puparia sampled (Fig. 1A). Those two methods of expressing parasitism are discussed by Petersen (1986a). Another 7% of the puparia, when dissected, were found to contain parasites. Therefore, at least 44.5% of the puparia were killed by parasites. It was also evident that some fly pupae were killed by parasites stinging the pupa and not producing a parasite, but an accurate number could not be determined. Legner (1979) reported 28-37% fly pupae killed by parasitism without parasites emerging from the puparia. If that rate of pupal kill occurs naturally (28-37% fly kill plus 37.5% parasite emergence), we have an ideal natural system to work with to obtain fly control. The actual cause of death of the unemerged pupae we dissected could not be determined.

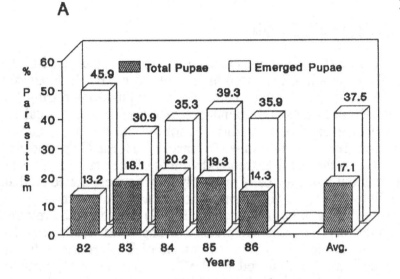

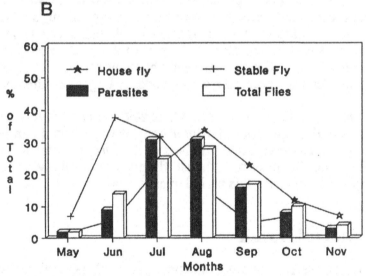

Figure 1. Parasites and flies collected from confined cattle feedlots in southwest Kansas. A. Parasite emergence from 69,751 fly puparia expressed as percent of total pupae collected and as percentage of emerged insects. B. Relative monthly abundance of fly puparia and total parasites emerging from all fly puparia collected.

Seasonal Population Trends

Stable fly populations peaked in June and decreased as the house fly populations increased in late July (Fig. 1B). The parasite populations lagged behind the stable fly populations during the spring. Parasite seasonal abundance was similar to the total fly abundance, even though it did not follow closely the seasonal population trends of individual fly species. House fly abundance was closer to the parasite trends than that of the stable fly, partly because about twice as many house fly puparia were collected than stable fly puparia. *S. nigroaenea* showed a definite summer peak occurring during July and August, (Fig. 2) matching the relative abundance of flies, then decreased along with decreasing fly populations in the fall. The low numbers of *S. nigroaenea* during May may have been caused by the low numbers of fly puparia collected, suggesting that releases of large numbers of *S. nigroaenea* in May might offer a good chance of reducing the early season stable fly populations. Early season releases may require supplying hosts for parasite development, as suggested by Petersen (1986b). *S. cameroni* was distributed more evenly during the summer than *S. nigroaenea* (Fig. 2). Similar seasonal abundance trends for fly and parasite species was reported from Uganda by Legner & Greathead (1969).

The monthly abundance of *M. zaraptor* was relatively even during July and August, but *M. raptor* had a definite July, (summer) peak (Fig. 3). However, low numbers of *M. zaraptor* parasitizing stable flies and the absence of house fly puparia during May-June make reported values for this period questionable.

Release Rates and Methods

A release rate of four parasites per week per animal was recommended by commercial distributors until recently, when some suppliers changed to twenty parasites per animal per week. The commercial rate mentioned by Petersen et al. (1983) was forty per animal per week. They released over 100 parasites per animal per week as did Stage & Petersen (1981) and did not see a reduction in fly numbers. Their release rates were low compared to the releases of 80-300 parasites per animal per week of Morgan et al. (1976) and Morgan (1980), who reported up to 100% control. The low release rates employed may be part of the reason why we have not yet

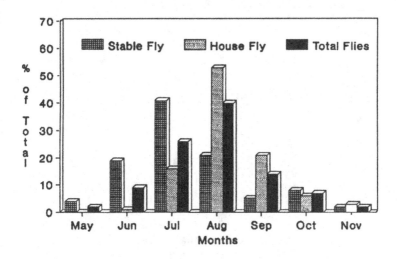

Spalangia cameroni

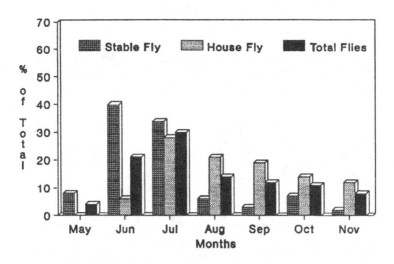

Figure 2. Average monthly occurrence of *S. nigroaenea* and *S. cameroni* emerging from stable fly, house fly, and total puparia of both fly species collected in southwest Kansas cattle feedlots during 1982-1986.

36

Muscidifurax zaraptor

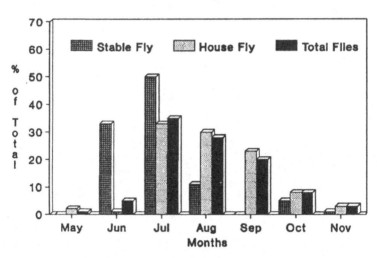

Muscidifurax raptor

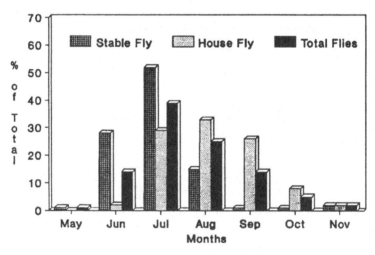

Figure 3. Average monthly occurrence of *M. zaraptor* and *M. raptor* emerging from stable fly, house fly, and total puparia of both fly species collected in southwest Kansas cattle feedlots during 1982-1986.

demonstrated fly control in cattle feedlots in the high plains. Much research work remains on this subject.

The method of releasing parasites is very much in question. Some firms simply scatter the parasites (parasitized pupae) on the ground where they are subject to predators and the hot sun. Petersen (personal communication) suggests that the environmental conditions around the released parasites may be critical for survival of active parasites. Antolin & Williams (1989) reported that feeding on pupal haemolymph by newly emerged female parasites increased reproductive potential. Unfed parasites absorbed eggs for energy, reducing the potential number that would be laid. Legner & Gerling (1967) reported that hosts are required for young adult parasites for maximum longevity and oviposition. Therefore, the parasite release rates and conditions (host food availability) may need to be changed to improve longevity and to facilitate efficient parasitism by mass-released parasites.

Parasite Species for High Plains Cattle Feedlots

Selection of the parasite species to release must be based on the fly species to be controlled. The parasite species for stable fly control appears to be *S. nigroaenea*, since it was the predominant parasite retrieved from stable fly puparia in Kansas, as well as Florida (Greene et al. 1989). *S. nigroaenea* was also collected from naturally-occurring puparia during parasite release studies by Stage & Petersen (1981) and Petersen et al. (1983) in Nebraska. As the fly species changes to house flies during July, it might be appropriate to switch to *M. zaraptor*. *M. zaraptor* was retrieved predominantly from house fly puparia (92%) and was present when house flies were abundant during July and August. *M. zaraptor* is more economical to mass rear than *S. nigroaenea*. Early seasonal abundance of *S. cameroni* during May-June (Fig. 2) suggests that it may be a better species than *S. nigroaenea* for early season stable fly control. However, in those feedlots where *S. cameroni* was mass released, greater numbers of *S. nigroaenea* emerged from collected puparia (unpublished data). *M. raptor* emerged from both fly species and may be a good candidate parasite for mass release, even though it contributed only 21% of the *Muscidifurax* collected from cattle feedlots.

Economics of Parasite Releases for Fly Control

Cost of controlling flies in cattle feedlots is of concern to feedlot operators. If parasites cost more than insecticides, operators would have little interest, except in regard to environmental contamination. The current advertised rates of four to twenty parasites per animal per week cost ten to forty-four cents per animal. Unfortunately, these numbers have not been shown to impact fly populations. Increasing release rates up to 100 parasites per animal per week, during the eight week stable fly season, could cost $2.00 per animal. Looking at the return to the producer and based on the estimates of two tenths pound loss in gain per day from stable fly feeding (Campbell et al. 1977), the loss would be 12 lb during the 60 day fly feeding period. Those figures equate to $7.20 loss per animal using 60 cents per lb sales price. By investing $2.00 for 100 parasites per animal per week, when stable fly populations are highest, the return would be $3.60 for each dollar invested.

Release of *S. nigroaenea*

S. nigroaenea was chosen as the parasite for mass release to control stable flies in a cattle feedlot during 1987, based on its relative abundance in stable fly puparia. The goal was to release fifty parasites per animal per week, but the low emergence rate (32% of mass reared *S. nigroaenea*) resulted in twenty-nine per animal per week being released. The results were encouraging, with the stable fly population peak in June being less than half the initial peak during 1985, a climatically similar year, and occurring about one month later (Fig. 4). The lack of reduction in late August can not be explained through either greater rainfall or reduced parasite release numbers. The parasites emerging from pupae collected in the feedlot during July included few *S. nigroaenea*. Before we judge the value of *S. nigroaenea* for mass release to control stable flies, additional releases will need to be made.

SUMMARY

Parasitism levels of filth fly pupae at cattle feedlots in the high plains were relatively high over a five year period, with 37.5% of the emergence from collected pupae being parasites. Two Pteromalidae genera, *Muscidifurax* and *Spalangia*, dominated the

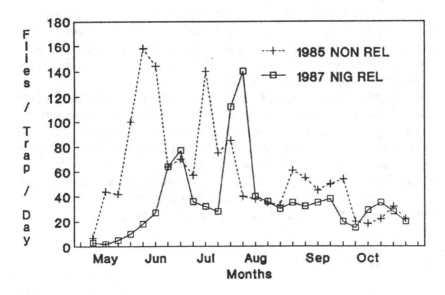

Figure 4. Adult stable flies caught on Alsynite traps at a cattle feedlot with similar fly populations and weather conditions in southwest Kansas during 1985 and 1987, with no parasites released during 1985 and *S. nigroaenea* released during 1987.

taxa recovered. Each genus had two predominant species, *M. raptor* and *M. zaraptor* and *S. cameroni* and *S. nigroaenea*, respectively. *M. zaraptor* and *S. nigroaenea* each contributed about 75% of the specimens for their genus, making them the logical species for mass releases. *M. zaraptor* is commercially available, but only 8% of the parasites recovered from stable fly puparia were *M. zaraptor*. *S. nigroaenea* appears to be the species adapted to cattle feedlot conditions in the high plains and should be considered for use to control stable flies. *M. zaraptor* and *S. nigroaenea* appear to be a good combination for house fly and stable fly control. *S. nigroaenea* populations peak during the summer, which fits the peak fly season; therefore, mass releases should be made each spring to impact the developing stable fly populations. The economic return from parasite releases for stable fly control looks favorable, with a possible $3.60 return for each dollar invested for parasites at a release rate of 100 parasites per animal.

Preliminary releases of *S. nigroaenea* in a cattle feedlot reduced the early season stable fly populations and delayed fly buildup by three weeks. A second fly peak in August was delayed three weeks, but the numbers of flies were not reduced. Parasite releases in cattle feedlots show promise for control of stable flies, provided several questions relative to parasite release efficiency can be answered. Repeated fly control using efficient and effective parasite release rates will be developed in the near future.

REFERENCES CITED

Axtell, R. C. 1981. Use of predators and parasites in filth fly IPM programs in poultry housing. In Proc. of Workshop, Status of Biological Control of Filth Breeding Flies. Gainesville, Fla. Feb. 3-5, 1981. pp. 26-43.

Antolin, M. F., & R. L. Williams. 1989. Host feeding and egg production in *Muscidifurax zaraptor* (Hymenoptera: Pteromalidae). Fla. Entomol. 72: 129-134.

Campbell, J. B., R. G. White, J. E. Wright, R. Crookshank, & D. C. Clanton. 1977. Effects of stable flies on weight gains and feed efficiency of calves on growing or finishing rations. Jour. Econ. Entomol. 70: 592-594.

Greene, G. L. 1989. Seasonal population trends of adult stable flies. Entomol. Soc. Amer. Misc. Pub. 74: 12-17.

Greene, G. L., J. A. Hogsette, & R. S. Patterson. 1989. Parasites that attack stable fly and house fly (Diptera: Muscidae) puparia during the winter on dairies in northwestern Florida. J. Econ. Entomol. 82: 412-415.

Legner, E. F. 1977. Temperature, humidity and depth of habitat influencing host destruction and fecundity of muscoid fly parasites. Entomophaga 22: 199-206.

_____. 1979. The relationship between host destruction and parasite reproductive potential in *Muscidifurax raptor, M. zaraptor*, and *Spalangia endius* (Chalcidoidea: Pteromalidae). Entomophaga 24: 145-152.

Legner, E. F., & H. W. Brydon. 1966. Suppression of dung-inhabiting fly populations by pupal parasites. Ann. Entomol. Soc. Amer. 59:638-651.

Legner, E. F., & D. Gerling. 1967. Host-feeding and oviposition on *Musca domestica* by *Spalangia cameroni, Nasonia vitripennis*, and *Muscidifurax raptor* (Hymenoptera: Pteromalidae) influences their longevity and fecundity. Ann. Entomol. Soc. Amer. 60: 678-691.

Legner, E. F., & D. J. Greathead. 1969. Parasitism of pupae in east African populations of *Musca domestica* and *Stomoxys calcitrans*. Ann. Entomol. Soc. Amer. 62: 128-133.

Legner, E. F., & G. S. Olton. 1968. Activity of parasites from Diptera: *Musca domestica, stomoxys calcitrans*, and species of *Fanna, Muscina*, and *Ophyra* II at sites in the eastern hemisphere and Pacific area. Ann. Entomol. Soc. Amer. 61: 1306-1314.

_____. 1970. Worldwide survey and comparison of adult predator and scavenger insect populations associated with domestic animal manure where livestock is artificially congregated. Hilgardia 40: 225-265.

_____. 1971. Distribution and relative abundance of Dipterous pupae and their parasitoids in accumulations of domestic animal manure in the southwestern United States. Hilgardia 40: 505-535.

Legner, E. F., E. C. Bay, & E. B. White. 1967. Activity of parasites from Diptera *Musca domestica, Stomoxys calcitrans, Fannia canicularis*, and *F. femoralis* at sites in the western Hemisphere. Ann. Entomol. Soc. Amer. 60: 462-468.

Meyer, J. A., & J. J. Petersen. 1982. Sampling stable fly and house fly pupal parasites on beef feedlots and dairies in Eastern Nebraska. Southwest Entomol. 7: 119-123.

Morgan, P.B. 1980. Sustained releases of *Spalangia endius* Walker (Hymenoptera: Pteromalidae) for control of *Musca domestica* L. and *Stomoxys calcitrans* (L.) (Diptera: Muscidae). J. Kans. Entomol. Soc. 53: 367-372.

Morgan, P. B., & R. S. Patterson. 1977. Sustained releases of *Spalangia endius* to parasitize field populations of three species of filth breeding flies. J. Econ. Entomol. 70: 450-452.

Morgan, P. B., R. S. Patterson, & G. C. LaBreque. 1976. Controlling house flies at a dairy installation by releasing a protelean parasitoid, *Spalangia endius* (Hymenoptera: Pteromalidae). J. Ga. Entomol. Soc. 11: 39-43.

Petersen, J. J. 1986a. Evaluating the impact of pteromalid parasites on filth fly populations associated with confined livestock installations. Entomol. Soc. Amer. Misc. Pub. 61: 52-56.

_____. 1986b. Augmentation of early season releases of filth fly (Diptera: Muscidae) Parasites (Hymenoptera: Pteromalidae) with freeze-killed hosts. Environ. Entomol. 15: 590-593.

Petersen, J. J., and J. A. Meyer. 1983. Host preference and seasonal distribution of pteromalid parasites (Hymenoptera: Pteromalidae) of stable flies and house flies (Diptera: Muscidae) associated with confined livestock in eastern Nebraska. Environ. Entomol. 12: 567-571.

Petersen, J. J., J. A. Meyer, D. A. Stage, & P. B. Morgan. 1983. Evaluation of sequential releases of *Spalangia endius* (Hymenoptera: Pteromalidae) for control of house flies and stable flies (Diptera: Muscidae) associated with confined livestock in eastern Nebraska. Jour. Econ. Entomol. 76: 283-286.

Rutz, D. A., & R. C. Axtell. 1979. Sustained releases of *Muscidifurax raptor* (Hymenoptera: Pteromalidae) for house fly (*Musca domestica*) control in two types of caged-layer poultry houses. Environ. Entomol. 8: 1105-1110.

Stage, D. A., & J. J. Petersen. 1981. Mass release of pupal parasites for control of stable flies and house flies in confined feedlots in Nebraska. In Proc. of Workshop, Status of Biological Control of Filth Breeding Flies. Gainesville, Fla. Feb. 4-5, 1981. pp. 52-58.

4. Biological Control as a Component of Poultry Integrated Pest Management

Jeffery A. Meyer

Livestock and poultry production in the United States is an important agricultural enterprise, with cash receipts totalling about $50 billion per year. Direct losses associated with arthropods attacking livestock and poultry are very substantial, estimated to be four billion dollars per year (Anon. 1979, Steelman 1976).

In addition to those arthropods inflicting direct losses to the livestock and poultry industries, there are other groups of arthropods, particularly filth breeding flies, that inflict substantial indirect losses. Examples of these losses would include fines and other fees associated with regulatory inspections, insecticide and equipment costs incurred to keep immature and adult fly populations below nuisance thresholds, and court and legal fees incurred through criminal and civil actions against livestock and poultry facilities that maintain excessive fly numbers. The extent of these losses has not been estimated, probably due to the difficulty in identifying and tabulating such types of data in the various regions of the country that intensively produce livestock and poultry commodities.

Of the confined livestock and poultry production systems currently utilized in the United States, indoor systems seem to have the greatest fly problem (Axtell 1986). Of these indoor systems, egg production facilities have probably received the greatest number of nuisance fly complaints. The manure management problems associated with the popular high-density housing systems have resulted in fly breeding conditions of great magnitude (Axtell 1986). Urbanization in the face of some of the intense egg production areas has increased the potential importance of dispersing fly populations. In many of these areas severe community relations problems have developed. The results of too many of these confrontations has been some type of monetary settlement (Low 1987).

43

The purpose of this paper is not to provide a detailed review of the scientific literature pertaining to biological control of poultry pests because comprehensive reviews have been recently published (Axtell & Rutz 1986). The intent of this discussion is to provide an opinion on the current role and future potential of biological control as a component of poultry IPM in the United States, as influenced by several different factors that may govern its use.

PRINCIPLES OF BIOLOGICAL CONTROL IN POULTRY IPM

The same integrated pest management (IPM) principles generally apply to any given agricultural commodity, but the importance of the different components can vary a great deal depending on the particular system. One of the key components of any IPM program is some type of meaningful economic threshold, whether it is based on pest damage, pest density, etc. (Stern 1973). Economic thresholds for nuisance flies on egg production facilities would have to be developed on a site-by-site basis. Factors influencing the development of economic thresholds for nuisance flies on caged-layer facilities would include the proximity of urban/suburban areas to the egg facility and the sensitivity of these areas to flies. Regulatory agencies, in effect, would determine the economic threshold of flies that could be tolerated on a given facility. Heavily urbanized areas would likely have more involvement of regulatory agencies, with the fly threshold being lower than that of a facility in a less urbanized area.

Current methods of managing and housing caged-layers offer potential in manipulating manure management to favor enhancement of natural enemies of filth-breeding flies. Modern poultry production uses high density, confinement systems which have provided conditions for development of potentially large populations of flies. It is apparent, therefore, that fly management is directly dependent upon manure management as an essential component in any integrated pest management program for poultry production systems. Legner et al. (1966) theorized that, under ideal conditions in an accumulated manure ecosystem, natural controls (biotic and abiotic) account for approximately 98% mortality of the immature fly stages. These high levels of mortality associated with biotic factors are achieved through habitat stability, which influences a diverse group of beneficial predatory and scavenger arthropods (Legner and Dietrick 1974).

Cultural control of flies through manure management can also include rapid removal and disposal of the manure, before the flies can complete a generation. The current urbanization trend that began in California during the 1940's has continued to the present. California's caged-layer industry was aware of the implications of this rapid urbanization and was beginning to feel its impact by 1963 (Anon. 1964). In response to the inevitable coexistence of the egg industry with the growing human population, the University of California initiated the Agricultural Sanitation Program (Loomis 1964). One of the results of this program was the development of various frequent manure removal systems, which were designed to remove poultry manure from the house at a frequent enough interval to prevent successful fly development (Fairbank 1964, Bell et al. 1965). The concepts were workable and reliable, and are still in practice today.

One important implication resulting from the development of these frequent manure removal techniques was that biological control through habitat stability (Legner and Dietrick 1974) was not necessarily a reliable option in the face of urbanization. Some caged-layer operators still utilize manure buildup systems, however. The choice of the particular method of manure management is definitely influenced by urbanization pressure and the preference of the respective county regulatory agency.

There are two general approaches regarding biological control of filth breeding flies in accumulated poultry manure. One deals with the enhancement of natural enemies through manure management and habitat stability (Legner and Dietrick 1974). The second theory involves augmentation or inundation of the developmental habitat with commercial fly parasites. Axtell and Rutz (1986) summarize the current literature regarding parasite releases and their relative success in controlling filth fly development. Many of the demonstrated successes of inundative or sustained releases of fly parasites were through controlled scientific studies (Axtell and Rutz 1986), with known parasite strains and high levels of quality control. These early successes initiated commercialization of the technology, with many insectaries selling fly parasites throughout North America (Table 1). Data are not available that document similar successes of controlling fly populations with commercial sources of fly parasites.

Whichever biological control principle is practiced, it must be remembered that biological control, in an ecological sense, can be defined as the regulation by natural enemies of another organism's

Table 1. North American sources of fly parasites.[1]

State/Province	# of Insectaries
California	11
Texas	2
Michigan	1
Indiana	1
Oregon	1
Virginia	1
Ohio	1
New York	1
South Dakota	1
Iowa	1
Washington	1
British Columbia	1
Ontario	1

[1] Source: Bezark, L. G. 1989. Suppliers of beneficial organisms in North America. Calif. Dept. of Food and Ag., BC 89-1.

population density at a lower average than would otherwise occur (DeBach 1974). Fluctuations about the mean are expected and may either be marked or essentially unnoticeable. By definition, natural enemies are dependent on the host's density. Density dependence operates by means of an intensification of the mortality-causing actions of enemies as the host or prey population increases and a relaxation of these actions as the host population density falls, so that host population increase beyond a characteristic high is prevented and decrease to extinction is prevented. Such reciprocal interaction results in the achievement of a typical average population density in a given habitat or area (DeBach 1974). An important factor to be considered when developing an IPM program for a particular caged-layer facility is the economic threshold of flies that must be maintained and the ability of a biological control program to maintain it below that level. Poultry IPM deviates from other agricultural IPM programs at this point, in that periodic fluctuations of the mean pest density above the economic threshold may not be tolerable in a poultry system (depending upon the urbanization pressure) where such fluctuations might be tolerated in an agricultural system.

INFLUENCE OF REGULATORY AGENCIES

The political structure of a given state and/or county can influence the potential impact of biological control in a poultry IPM system. In many states flies are considered public health threats and are governed by public health laws and regulations. Many of these laws are very strict at the state level, with some counties in California enacting their own fly control ordinances for poultry facilities that are even more restrictive than state law. With these facts it is apparent that fly control on poultry facilities in urbanized areas must be efficient and consistent, or producers could face severe criminal or civil penalties.

To better illustrate the role of poultry and human demographics in influencing the role of biological control in poultry IPM, the three county/metropolitan areas with the largest level of egg production were selected from the three top poultry producing states (California, Indiana, and Pennsylvania) (Unit. States Dept. Ag. 1988). Human and laying hen population densities were determined for each area (U.S. Bureau of the Census 1987a, b, c, d, e, f) and are shown in Table 2.

California has the largest number of caged-laying hens in the United States, with approximately 30% (15,653,152) of the total number clumped into the Riverside/San Bernardino metropolitan area in southern California, along with 6.8% (2,001,100) of the state's human population. The Pennsylvania egg industry is much smaller than California's, but has a similar clumped pattern. Over 46% (11,426,855) of Pennsylvania's caged laying hens are located in Lancaster Co., with a human population of 371,700 (3.1% of state's total). Indiana's caged-laying hen population is approximately the same as Pennsylvania's, but the largest egg producing area is Dubois Co., with 2,726,911 hens, or 15.0% of the state's total. Dubois Co., however, only contains 0.6% of Indiana's total population.

By itself, this data is not informative as to the urbanization pressure placed on the caged-layer industry in the respective states. The stability and nature of the human population in the respective counties would be more indicative of the actual pressure placed on

Table 2. Populations of caged-laying hens (1987) and humans (1986) in the three states with the largest number of caged-laying hens.[1]

State (Co./Metro Area)	# Laying Hens	# Humans
California	40,712,228	26,981,000
Riverside/San Bernardino Cos.	15,653,152	2,001,100
Indiana	23,215,449	5,504,000
Dubois Co.	3,345,782	36,000
Pennsylvania	21,608,619	11,888,000
Lancaster Co.	11,426,855	393,500

[1] Source: U. S. Bureau of the Census, census of agriculture and current population reports.

the respective caged-layer industry. Table 3 lists the top ten counties in the United States (irrespective of state) with the largest population growth between 1980 and 1986. It is obvious that southern California continues to have a rapidly expanding human population, since five of the top ten counties are in southern California, including Riverside and San Bernardino. In comparison, the population of Lancaster Co. only increased by approximately 24,000 from 1980 to 1985, while Dubois Co. only increased by 1,500 over the same time period (U.S. Bureau of the Census 1987e).

The burgeoning egg industry in southern California is comingled with a an equally burgeoning human population, which has created a classic urban/agricultural interface. The main result of this association has been greater and greater demands for more effective and immediate fly control. These demands are currently not being met and are not likely to be satisfied by biological control. This is not to say that biological control will not be included as a control technology for poultry IPM programs in southern California, but it is not likely to remain as large a component as it might in Lancaster Co. or Dubois Co.

Table 3. Top ten U. S. counties with largest population growth between 1980 and 1986.[1]

County (State)	Population (1986)	Population Growth (1980-1986)
Los Angeles (CA)	8,295,900	818,700
Maricopa (AZ)	1,816,700	391,000
Harris (TX)	2,798,300	388,800
San Diego (CA)	2,201,300	339,500
Dallas (TX)	1,833,100	276,700
San Bernardino (CA)	1,139,100	244,100
Tarrant (TX)	1,043,600	240,700
Orange (CA)	2,127,000	233,900
Riverside (CA)	818,400	198,800
Bexar (TX)	1,170,000	181,100

[1] Source: U. S. Bureau of the Census, County and City Data Book, 1988. U.S. Government Printing Office, Washington D. C.

INFLUENCE OF BIOLOGICAL CONTROL ON INSECTICIDE USE

Reviews of major North American entomological journals revealed a total of eighty-six scientific articles published from 1968-1987, which dealt with biological control of filth flies on poultry facilities. If indeed this basic research is being implemented at the field level, a gradual decrease in the amount of insecticide used on poultry facilities would be expected. Table 4 does show a decrease in insecticide use on livestock (including poultry) from 1966-1976, which might be attributed to implementation of biological control and IPM. However, during that time period poultry production was beginning to become more intensified and systematic. The number of farms with greater than 20,000 hens in California, Indiana and Pennsylvania generally declined from 1976 to 1987, with a corresponding increase in the number of hens (Fig. 1). The implication from these figures is that fewer buildings were beginning to house more hens. With the number of poultry houses decreasing, a corre-

50

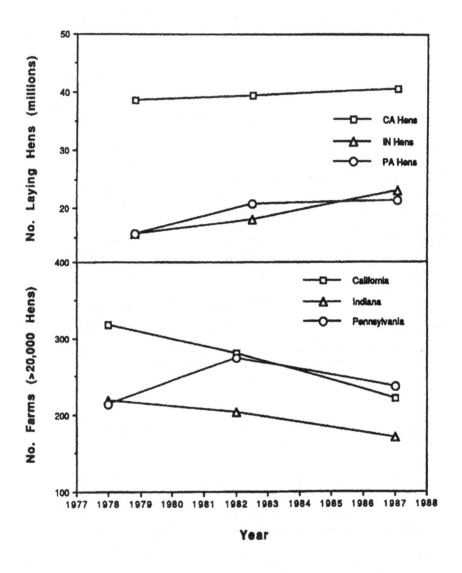

Figure 1. Temporal trends in number of caged-laying hens and number of farms housing greater than 20,000 hens, in California, Indiana and Pennsylvania.

Table 4. Farm use of pesticides on all crops and livestock, 1966 and 1976.[1]

| Type of Pesticide | Quantity of Active Ingredient (mil. lbs.) | | |
	1966	1976	% Change
Crop Pesticides			
Herbicides	117.0	394.0	237.0
Insecticides	138.0	162.0	17.0
Fungicides	33.0	43.0	30.0
Other	41.0	50.0	22.0
Total	329.0	649.0	94.0
Livestock Insecticides			
Beef	--[2]	7.5	
Dairy		2.0	
Hogs		0.7	
Poultry		0.3	
Other		0.5	
All Livestock	12.0	11.0	-8.3

[1] Source: Eichers, T. R. 1981. Use of pesticides by farmers. In D. Pimentel ed., Handbook of Pest Management in Agriculture, Volume 2. 501 pp.

[2] Insecticide use data for individual commodities during 1966 is not available.

sponding decrease in insecticide use might also be expected. The concept can be compared to a reduced amount of acreage of a crop.

Another confounding factor in this relationship would be the relatively small amount of pesticide utilized by the poultry industry, as compared to many crops. The measurable impact of implementing biological control and IPM through reductions in insecticide use could be demonstrated more easily in crops such as corn or cotton, which receive much larger levels of insecticide (Table 4). It would be difficult to detect a decrease in pesticide usage and attribute it to biological control, when the absolute amount of insecticide actually used is small. Also, the data for poultry in Table 4 includes insecticides that are applied for ectoparasite control on poultry and no biological control strategies exist for these pests.

It is unfortunate that pesticide use information is not available for livestock (and poultry) after 1976, because the declining pattern is likely to have continued to the present. One reason for this possible decline could be continued intensification of poultry housing, as mentioned previously, but the decline might also be indicative of problems associated with developing and registering new insecticides for the poultry industry. In terms of pesticide use, poultry could be considered a minor use commodity, and chemical companies may consider not registering new insecticides or not re-registering old insecticides. Development and registration of a new synthetic pesticide can take from eight to ten years and cost $20-40 million (Knight and Norton 1989), while re-registration of existing pesticides costs approximately $150,000. Since poultry utilizes small amounts of insecticide in relation to other agricultural commodities (Table 4), pesticide companies are probably reluctant to develop new compounds for the poultry market. According to Knight and Norton (1989), the pesticide industry ignores pest problems of minor crops. This problem is particularly acute in California, where registration requirements are generally higher than those of the Environmental Protection Agency. The ultimate result of the reduced availability of insecticides for the poultry industry may be greater awareness and implementation of biological control and IPM principles.

The other result may be an overuse and abuse of available insecticides (Meyer et al. in press). The results of this abuse have been documented repeatedly over the past twenty-five years in the United States, with filth flies developing metabolic resistance to a wide array of insecticides (Georghiou and Bowen 1966, Horton et al. 1985, Meyer et al. 1987, Meyer et al. in press). An analogy to this problem in vegetable crops was discussed by Trumble and Parrella (1987), when they reviewed California's policies for registration of new pesticides. The California Food and Agriculture Code (part c of section 12825 in article 4) prohibits registration of additional chemicals if a "feasible alternative" is currently registered. This law was enacted to reduce pesticide use and facilitate residue monitoring. However, this law exacerbates the problem of resistance and also reduces the potential for exploiting biological control agents by limiting the chemicals available for an integrated pest management program (Trumble and Parrella 1987).

An excellent example of this problem is the failure of the California Department of Food and Agriculture to register cyromazine (Larvadex®, CIBA-GEIGY Corp.) as a feed-through larvicide for fly control on caged-layer facilities. Cyromazine is highly efficacious and has a completely different mode of action than other larvicides registered for filth fly control, and is also compatible with predatory Coleoptera in the manure (Axtell and Edwards 1983, Meyer et al. 1983). The addition of cyromazine to the available fly control chemicals would not only have given producers a highly efficacious insecticide that may have prolonged their existence in urbanized surroundings, but would have allowed for greater flexibility in managing resistance development. This ruling not only increased chances for flies developing resistance to registered chemicals, but indirectly limited the use of biological control in highly urbanized areas of California.

SUMMARY AND CONCLUSIONS

A variety of factors currently influence the role of biological control in poultry IPM and these factors do not exert their influence uniformly across the United States. Perhaps the most important factor influencing the development and implementation of biological control for poultry pests is urbanization, or the severity of urban/agricultural interface. Poultry production in heavily urbanized areas will be subject to much greater scrutiny in terms of control of filth flies, and much greater levels of fly control will be necessary. Regulatory agencies will probably have much larger roles in dictating fly control standards on poultry facilities. Their fly control requirements will be dictated by complaints registered by neighbors of poultry facilities.

Entomologists have worked for years on many different aspects of biological control of filth flies, with the majority of the research directed at pupal parasites. As previously mentioned, significant natural mortality occurs in the immature stages of the house fly (Legner 1966), so biological control opportunities directed toward these stages may be limited. Future filth fly biological control research should include a greater emphasis on adult flies. Some promising results have been recently published regarding the effect of *Entomophthora muscae* (Cohn) Fresenius on adult house flies (Mullens et al. 1987). Biological control of the adult flies would not only improve control of the pestiferous stage, but would exploit different control mechanisms. Natural control of adult flies would also reduce the probability of the development of insecticide

54

resistance, since resistance genes are generally selected in adult populations.

One exception to this line of thinking might be to re-evaluate classical biological control of flies. Legner and co-workers made great strides in identifying and importing exotic strains of parasites from various locales into the United States, for evaluation against filth breeding flies. Currently the United States Department of Agriculture maintains a facility in Europe to support exploratory activities for biological control organisms of veterinary pests, including filth flies (Hoyer 1986).

Biological control is not likely to become a larger component of poultry IPM in the near future, for the reasons previously discussed. For biological control to become a larger and more reliable component of poultry IPM, much more research is necessary. The amount of biological control/IPM research probably will not increase in the near future, because there is no impetus on the part of state or federal granting agencies to provide funding. That is, implementing more biological control would have no tremendous impact in decreasing the amount of insecticide used to control flies. If the poultry industry, particularly the caged-layer industry, is willing to support long-term research projects in this area, then the emphasis is likely to change.

REFERENCES CITED

Anonymous. 1964. Urbanization's impact on California's poultry industry. Pacific Poultryman, Feb., 1964.

Anonymous. 1979. Proc. Workshop on Livestock Pest Management: To Assess National Research and Extension Needs for Integrated Pest Management of Insects, Ticks, and Mites Affecting Livestock and Poultry. Kansas State Univ., Manhattan, 322 p.

Axtell, R. C. 1986. Fly Control in Confined Livestock and Poultry Production. Technical Monograph, CIBA-GEIGY Corp., Greensboro, NC, 59 pp.

Axtell, R. C. and T. D. Edwards. 1983. Efficacy and nontarget effects of Larvadex® as a feed additive for controlling house flies in caged-layer poultry manure. Poult. Sci. 62: 2371-2377.

Axtell, R. C. and D. A. Rutz. 1986. Role of parasites and predators as biological fly control agents in poultry production facilities, pp. 88-100. In: Biological Control of Muscoid Flies. Miscellaneous Publication No. 61, Entomological Society of America, College Park, Maryland.

Bell, D. D., R. G. Curley, and E. C. Loomis. 1965. Poultry manure removal systems. Univ. Calif. Agric. Ext. Ser., Pub. AXT-189.

Bokhari, S. 1989. Avoid nuisance complaints. Poultry Digest, July 1989, pp. 334-335.

DeBach, P. 1974. Biological Control by Natural Enemies. Cambridge University Press, 1st ed. New York, New York.

Eichers, T. R. 1981. Use of pesticides by farmers, pp. 3-25. In D. Pimentel [ed.], Handbook of Pest Management in Agriculture, Volume II, CRC Press, Boca Raton, Florida.

Fairbank, W. C. 1964. Cage house design for manure removal. Univ. Calif. Ag. Ext. Ser. , Pub. AXT-110.

Georghiou, G. P. and W. R. Bowen. 1966. An analysis of house fly resistance to insecticides in California. J. Econ. Entomol. 59: 204-214.

Horton, D. L, D. C. Sheppard, M. P. Nolan, Jr., R. J. Ottens and J. A. Joyce. 1985. House fly (Diptera: Muscidae) resistance to permethrin in a Georgia caged-layer poultry operation. J. Agric. Entomol. 2: 196-199.

Hoyer, H. 1986. Survey of Europe and North Africa for parasitoids that attack filth flies, pp. 35-38. In: Biological Control of Muscoid Flies. Miscellaneous Publication No. 61, Entomological Society of America, College Park, Maryland.

Knight, A. L. and G. W. Norton. 1989. Economics of agricultural pesticide resistance in arthropods. Ann. Rev. Entomol. 34: 293-313.

Legner, E. F. and C. W. McCoy. 1966. The housefly, *Musca domestica* Linnaeus, as an exotic species in the western hemisphere incites biological control studies. Can. Entomol. 98: 243-248.

Legner, E. F. and E. I. Dietrick. 1974. Effectiveness of supervised control practices in lowering population densities of synanthropic flies on poultry ranches. Entomophaga 19: 467-478.

Loomis, E. C. 1964. Agricultural sanitation and the domestic fly program. Proc. Calif. Mosq. Cont. Assoc. 32: 34-36.

56

Low, C. 1987. What's 'unfair' in common scents. Insight, November 9, 1987, p. 56.

Meyer, J. A., W. F. Rooney, and B. A. Mullens. 1984. Effect of Larvadex® feed-through on cool-season development of filth flies and beneficial Coleoptera in poultry manure in southern California. Southwest. Entomol. 9: 52-57.

Meyer, J. A., G. P. Georghiou & M. K. Hawley. 1987. House fly (Diptera: Muscidae) resistance to permethrin on southern California dairies. J. Econ. Entomol. 80: 636-640.

Meyer, J. A., G. P. Georghiou, F. A. Bradley & H. Tran. In press. Filth fly resistance to pyrethrins associated with automated spray equipment in poultry houses. Poult. Sci.

Mullens, B. A., J. L. Rodriguez, and J. A. Meyer. 1987. An epizootiological study of *Entomophthora muscae* in muscoid fly populations on southern California poultry facilities, with emphasis on *Musca domestica*. Hilgardia 55: 1-41.

Steelman, C. D. 1976. Effects of external and internal arthropod parasites in domestic livestock production. Ann. Rev. Entomol. 21: 155-178.

Stern, V. M. 1973. Economic thresholds. Ann. Rev. Entomol. 18: 259-280.

Trumble, J. T. & M. P. Parrella. 1987. California law and the development of pesticide resistance. California Policy Seminar Research Report, University of California, Berkeley.

U. S. Bureau of the Census, County and City Data Book. 1988. U. S. Government Printing Office: 1988.

U. S. Bureau of the Census, Census of agriculture. 1987a. Geographic area series. Part 14, Indiana state and county data.

_____. 1987b. Geographic area series. Part 5, California state and county data.

_____. 1987c. Geographic area series. Part 38, Pennsylvania state and county data.

_____. 1987d. Current population reports. U. S. Census Bureau Series P-26, No. 85-CA-C. Estimates of the population of California counties and metropolitan areas: July 1, 1981, to 1985.

_____. 1987e. Current population reports. U. S. Census Bureau Series P-26, No. 85-IN-C. Estimates of the population of Indiana counties and metropolitan areas: July 1, 1981, to 1985.

U. S. Bureau of the Census, Census of agriculture. 1987f. Current population reports. U. S. Census Bureau Series P-26, No. 85-PA-C. Estimates of the population of Pennsylvania counties and metropolitan areas: July 1, 1981, to 1985.

U. S. Department of Agriculture. Agricultural Statistics. 1988. U. S. Government Printing Office, Washington: 1988.

5. Survey of House Fly Pupal Parasitoids on Dairy Farms in Maryland and New York

R. W. Miller and D. A. Rutz

Over the past five years scientists at the Livestock Insects Laboratory in Beltsville, Maryland, and the Department of Entomology at Cornell University have been involved in a cooperative research project to evaluate the potential of using hymenopterous pupal parasitoids as a part of an integrated house fly management program for use on dairy farms in the northeastern region of the United States. The initial phases of this study included surveys to determine the relative and seasonal abundance of naturally occurring pupal parasitoids on these farms in Maryland and New York. Results of these surveys provided direction for species to be mass reared and released as a component of an overall house fly management program.

Background on the types of farms used in these studies is described in a paper by Lazarus et al. (1989). Although the farms in Maryland and New York were generally similar in regard to the number of cows and the type of housing, there were some notable differences, primarily in the housing of calves and heifers. In Maryland, calves tended to be housed in outdoor facilities, i.e., homemade or purchased individual calf hutches. Also, Maryland heifers were generally kept on pasture during the summer months. In New York calves were usually housed indoors, often in barns converted for this use. Heifers were generally held in barns or open-sided sheds even during the summer months. These housing differences often meant that a larger proportion of the fly breeding took place inside of barns on farms in New York than on farms in Maryland.

Data on relative and seasonal abundance of house fly pupal parasitoids in the two states are presented in this paper. Preliminary findings on inundative releases of the parasitoid *Muscidifurax*

59

60

raptor, Girault and Sanders, on four farms in Maryland are also discussed.

MATERIALS AND METHODS

For the parasitoid surveys, ten screened bags, each measuring 10 x 12 cm and containing thirty, one-day-old house fly pupae, were placed around prime fly-breeding areas on each farm (Rutz and Axtell 1980). Approximately half of the bags were placed inside barns or sheds and half were placed in outdoor areas. Each week the bags were collected, taken to the laboratory, and held for flies and parasitoids to emerge and die. Flies and parasitoids were then counted and species of parasitoids were identified.

In Maryland these surveys were conducted on eight, eight, and eleven dairy farms during the years of 1983, 1984, and 1985, respectively. Surveys were conducted on eight, eight, and five dairy farms in New York for the same three years.

In 1984, during the last week of August and for four weeks in September, two parasitoid species, *M. raptor* and *Urolepis rufipes* (Ashmead), were released on a dairy farm in Maryland. Ca. 20,000 and 19,500 pupae parasitized by *M. raptor* and *U. rufipes,* respectively, were placed in screened release cages in a stanchion barn converted for calf pen housing and storage. A similar study was conducted in New York, and the results were published by Smith (1988).

In 1985, from the first of June until the middle of September, *M. raptor* were released on four dairy farms in Maryland (obtained from IPM Laboratories, Locke, New York). On two farms, ca. 13,500 parasitized house fly pupae were set out per week and on two other farms, ca. 4,500 parasitized pupae were set out. Pupae containing the parasitoids were placed in screened bags and hung on a wall of a barn in an area of high fly-breeding potential. One bag was used on the farms for the lower release rate and three bags were used on the higher release rate farms. Samples of the parasitized pupae to be released were held in the laboratory to determine the percentage of emergence and the sex ratio of the parasitoids.

RESULTS AND DISCUSSION

A total of ten parasitoid species were found on dairy farms in Maryland and New York over the three-year period. These included *M. raptor, Nasonia vitripennis* Walker, *Pachycrepoideus vindemmiae* (Rondani), *Phygadeuon fumator* Gravenhorst, *Spalangia cameroni* Perkins, *S. endius* Walker, *S. nigra* Latreille, *S. nigroaenea* Curtis, *Trichomalopsis dubius* (Ashmead), and *U. rufipes*.

In each of the three years, two to four times more of the sentinel pupae were parasitized in New York than in Maryland (Table 1). In both states, *M. raptor* accounted for a majority of the parasitism. A higher percentage of the pupae was parasitized by this species in Maryland than in New York. Whether or not the sampling procedure used biased the estimates of *M. raptor* is not known, however; Rutz (1986) discussed the possibility of this.

The second most abundant parasitoid sampled in Maryland was *S. cameroni*. In each of the three years, *M. raptor* and *S. cameroni* accounted for over 90% of the total parasitoids collected in Maryland. Other species sampled in Maryland accounted for 6% or less of the total parasitism. These species, listed in order of declining abundance, included *N. vitripennis, S. nigroaenea, P. vindemmiae*, and *S. endius*.

In New York, a lower percentage of the sentinel pupae were parasitized by *M. raptor* than in Maryland. However, more pupae were parasitized by other species. In 1984 and 1985, *U. rufipes* accounted for the second most parasitism, while in 1983 *S. cameroni* did. *U. rufipes*, which was not found in Maryland, was found on all eight farms surveyed in 1983 (Smith and Rutz 1985). A majority of this species was collected outdoors, primarily from manure pits and other accumulations of wet manure. Peterson et al. (1985) also found *U. rufipes* parasitizing muscoid fly pupae on a dairy farm in Nebraska.

Other species sampled in New York accounted for up to 15.7 percent of the total parasitism of the sentinel pupae. These species, listed in order of declining abundance, included *P. fumator, S. nigroaenea, P. vindemmiae, T. dubius, N. vitripennis*, and *S. nigra*. The occurrence of *T. dubius* on New York dairy farms was discussed by Hoebeke and Rutz (1988).

Table 1. Relative percentage of parasitization by indigenous parasitoids of laboratory-reared house fly pupae exposed in screened sentinel bags on dairy farms in Maryland and New York.

Species and Year	Relative Parasitization (%)											
	June		July		August		September		October		% of total	
	MD	NY	MD	NY	MD	NY	MD	NY	MD	NY	MD	NY
Muscidifurax raptor												
1983	—	—	—	—	78.6	62.3	82.6	47.4	82.0	81.8	80.6	63.1
1984	—	—	91.1	65.2	86.3	62.4	54.7	60.6	95.9	52.2	85.1	61.1
1985	56.5	0	83.2	38.8	91.6	60.5	93.9	50.7	—	—	87.8	51.7
Spalangia cameroni												
1983	—	—	—	—	10.9	20.4	14.8	16.8	17.4	7.7	13.5	14.4
1984	—	—	0	0.2	11.0	2.9	42.5	7.9	2.8	21.7	12.5	6.5
1985	17.6	0	0	0	8.1	10.3	4.1	24.9	—	—	6.1	13.4
Urolepis rufipes												
1983	—	—	—	—	0	8.1	0	20.2	0	4.4	0	11.8
1984	—	—	0	17.0	0	9.5	0	23.2	0	17.3	0	16.7
1985	0	0	0	53.0	0	13.5	0	11.7	—	—	0	22.1
All others												
1983	—	—	—	—	10.4	9.2	2.6	15.5	0.6	6.2	5.9	10.2
1984	—	—	8.9	17.5	2.7	25.3	2.8	8.3	1.3	8.7	2.4	15.7
1985	25.9	0	16.8	8.2	0.2	15.6	2.0	12.7	---	---	6.0	12.8
No. pupae parasitized												
1983	—	—	—	—	412	617	304	1092	161	941	877	2650
1984	—	—	56	1004	182	931	181	935	462	538	881	3408
1985	85	0	161	428	364	717	343	691	---	---	945	1837

Table 2. Percentage of seasonal abundance of the major parasitoids emerging from laboratory-reared house fly pupae exposed in screened sentinel bags on dairy farms in Maryland and New York.

| Species and Year | Seasonal Abundance (%) | | | | | | | | | | Total no. collected | |
| | June | | July | | August | | September | | October | | | |
	MD	NY	MD	NY	MD	NY	MD	NY	MD	NY	MD	NY
Muscidifurax raptor												
1983	—	—	—	—	45.8	23.0	35.5	31.0	18.7	46.1	707	1672
1984	—	—	6.8	31.3	20.9	27.8	13.2	27.5	59.1	13.4	750	2093
1985	5.8	0	16.1	17.5	39.3	45.7	38.8	36.8	—	—	830	950
Spalangia cameroni												
1983	—	—	—	—	38.1	33.0	38.1	48.2	23.7	18.8	118	382
1984	—	—	0	0.9	18.2	12.2	70.0	33.9	11.8	52.9	110	221
1985	25.9	0	0	0	50.0	30.1	24.1	69.9	—	—	58	246
Urolepis rufipes												
1983	—	—	—	—	0	16.0	0	70.8	0	13.1	0	312
1984	—	—	0	29.8	0	15.4	0	38.6	0	16.2	0	573
1985	0	0	0	56.0	0	24.0	0	20.0	---	---	0	406
% Parasitism												
1983	—	—	—	—	10.5	7.3	7.0	10.6	4.0	11.2		
1984	—	—	1.0	7.3	4.2	6.8	5.7	8.1	8.3	5.7		
1985	1.7	0	3.7	0.9	6.3	1.6	2.9	1.5	---	---		

A majority of the parasitism by the three predominant species occurred in August, September, and October; however, *U. rufipes* appeared to have a higher percentage of its seasonal abundance in July than did *M. raptor* and *S. cameroni* (Table 2). This confirms the work reported by Smith and Rutz (1985). The overall percentage of parasitism was higher in New York than in Maryland in 1983 and 1984. This would be expected because of the higher total number of parasitoids collected from the sentinel pupae. In 1985, there was a marked drop-off in overall parasitism in New York, and during each month parasitism was higher in Maryland than in New York.

In 1984, on the farm in Maryland where parasitoids were released in August and September, laboratory data showed that an average of 12,000 and 9,000 female *M. raptor* and *U. rufipes*, respectively, were released weekly. During September and October, the percentage of parasitism of sentinel pupae by *M. raptor* was high (up to 80 percent) at two of the three sites in the calf barn where the releases were made. No parasitism by *M. raptor* occurred at any of the other eight sites on the farm, including one in the release barn. Smith (1988) found that in New York, *M. raptor*, which were released inside the barn, parasitized pupae both inside and outside and achieved highest rates of parasitism in indoor straw bedding and in outdoor manure and silage.

In Maryland, no *U. rufipes* were found emerging from any of the sentinel pupae. This may have been because our releases were made inside the barn. Smith (1988) found in New York that *U. rufipes* did not attack house fly pupae inside the barn, and its highest rates of parasitism occurred in outdoor silage and manure. One would think that *U. rufipes* could survive under Maryland weather conditions; however, we have never found this species in sentinel pupae surveys conducted over a five-year period.

Results obtained from the farms on which *M. raptor* parasitoids were released in Maryland in 1985 are presented in Table 3. Laboratory trials showed that 23% of the parasitized pupae produced female parasitoids; therefore, the average number of female *M. raptor* released weekly was 1,050 for the lower release farms and 3,150 for the higher release farms. A higher proportion of the seasonal abundance of *M. raptor* occurred in June and July on the release farms than on the non-release farms (data in Table 3 compared with data in Table 2). An exception to this trend was noted for the higher release farms in June. The percentage of parasitism was also higher on the release farms than on the non-release farms in both June and July. This trend held for the higher release farms throughout the season. Average percentage of parasitism during the

Table 3. Percentage of seasonal abundance of major parasitoids emerging from laboratory-reared house fly pupae exposed in screened sentinel bags on dairy farms in Maryland on which *Muscidifurax raptor* parasitoids were released in 1985.

Species	June		July		August		September		Total No. Collected	
	LR[a]	HR[b]	LR	HR	LR	HR	LR	HR	LR	HR
Muscidifurax raptor	39.6	0.8	23.8	25.5	24.4	43.0	12.2	30.7	164	662
Nasonia vitripennis	0.9	67.6	92.8	0	3.6	32.4	2.7	0	111	367
Spalangia cameroni	0	0	0	0	32.1	0	67.9	100	28	25
% Parasitism	5.6	9.6	4.5	12.2	4.1	23.7	1.2	8.3		

a LR = lower release rate.
b HR = higher release rate.

summer was 3.6 percent, 3.8 percent, and 13.4 percent for the non-release, lower release, and higher release farms, respectively.

These studies have shown that *M. raptor* is the predominant indigenous house fly pupal parasitoid on dairy farms in Maryland and New York. This species appeared to be primarily responsible for the natural parasitism that varied between years, but was as high as 10% to 11% during some months. Peak parasitism generally occurred during the later part of the season, i.e., August to October.

Inundative releases of *M. raptor* early in the season in Maryland demonstrated that a higher proportion of the parasitism caused by *M. raptor* could be shifted to June and July. This finding suggests that the release of *M. raptor* could aid in controlling house fly and possibly stable fly populations when initial populations of these species are low. Additional work in this area has been conducted and will be reported in future papers.

REFERENCES CITED

Hoebeke, E. R., and D. A. Rutz. 1988. *Trichomalopsis dubius* (Ashmead) and *Dibrachys cavus* (Walker): newly discovered pupal parasitoids (Hymenoptera: Pteromalidae) of house flies and stable flies associated with livestock manure. Ann. Entomol. Soc. Am. 81:493-497.

Lazarus, W. F., D. A. Rutz, R. W. Miller, and D. A. Brown. 1989. Costs of existing and recommended manure management practices for house fly and stable fly (Diptera: Muscidae) control on dairy farms. J. Econ. Entomol. 82:1145-1151.

Peterson, J. J., D. R. Guzman, and B. M. Pawson. 1985. *Urolepis rufipes* (Hymenoptera: Pteromalidae), a new parasite record for filth flies (Diptera: Muscidae) in Nebraska, USA. J. Med. Entomol. 22:345.

Rutz, D. A. 1986. Parasitoid monitoring and impact evaluation in the development of filth fly biological control programs for poultry farms. In Biological control of Muscoid Flies, R. S. Patterson and D. A. Rutz, (eds.), Entomol. Soc. Am., College Park, pp. 45-51.

Rutz, D. A., and R. C. Axtell. 1980. House fly (*Musca domestica*) parasites (Hymenoptera: Pteromalidae) associated with poultry manure in North Carolina. Environ. Entomol. 9:175-180.

Smith, L. 1988. Dispersal behavior of two pteromalid parasitoids of house fly pupae in a dairy environment (Hymenoptera: Chalcidoidea). In Advances in Parasitic Hymenoptera Research, V. K. Gupta (ed.), E. J. Brill, Leiden, pp. 333-344.

Smith, L., and D. A. Rutz. 1985. The occurrence and biology of *Urolepis rufipes* (Hymenoptera: Pteromalidae), a parasitoid of house flies in New York dairies. Environ. Entomol. 14:365-369.

6. Efficiency of Target Formulations of Pesticides Plus Augmentative Releases of *Spalangia endius* Walker (Hymenoptera: Pteromalidae) to Suppress Populations of *Musca domestica* L. (Diptera: Muscidae) at Poultry Installations in the Southeastern United States

Philip B. Morgan and Richard S. Patterson

ABSTRACT

Augmentative releases of *Spalangia endius* Walker in conjunction with cyromazine against field populations of *Musca domestica* L. at caged layer installations were effective in controlling the *M. domestica* populations. Manure management is vital when using either biological control or chemical control or combinations of both against populations of *M. domestica*. Monitoring the third instar larval populations provided a reliable method for evaluating the effectiveness of the parasite releases and the cyromazine treatment.

KEY WORDS: *Spalangia endius, Musca domestica, Muscidifurax raptor, Spalangia cameroni, Spalangia nigroaenea, Stomoxys calcitrans*, parasite, cyromazine.

It has been demonstrated that augmentative releases of *Spalangia endius* Walker can be used to control field populations of *Musca domestica* L. and *Stomoxys calcitrans* (L.) in Florida at poultry (Morgan et al. 1975a,b, 1981a,b) and dairy (Morgan et al. 1976; Morgan & Patterson 1977) farms. The speed and effectiveness of cyromazine as a feed through to control field populations of *M.domestica* had been reported by Hall & Foehse (1980), Williams & Berry (1980), and Mulla & Axelrod (1983a,b). However, continued use of cyromazine has resulted in the wild house fly population developing resistance (Bloomcamp et al. 1987). Therefore, it was the object of this study to reduce the fly population with a two-week application of cyromazine, followed by parasite releases to maintain a high level of fly suppression. The research was conducted at four farms in south Georgia and north Florida. We report herein the results of this study.

MATERIALS AND METHODS

This study, which was initiated in June and terminated in October when the fly season ended, was conducted in an isolated area along the Florida-Georgia border at four small caged layer farms, each accommodating 20,000-30,000 birds in open-sided California style houses. The manure accumulated on the soil directly beneath the cages and was spread on the fields at twelve- to eighteen-month intervals. The third-instar larval and the one-, two, three-, and four-day-old house fly pupal populations at each farm were determined, using the model developed by Morgan et al. (1981a,b). Wild house fly pupae (100-200) were collected at each site and examined for the presence of parasite eggs, larvae, pupae, and adult parasites (Morgan et al. 1981b). The parasites that emerged were identified by the senior author using the taxonomic keys of Boucek (1963) and Peck et al. (1964). The parasites used for the augmentative release were a Florida strain of *Spalangia endius* Walker. The colony had been in culture for 126 generations and was maintained by the method described by Morgan (1981).

The following treatment schedules were initiated in June 1983. At the first farm, the insect growth regulator cyromazine, at a concentration of 0.05%, was included in the poultry feed for two weeks. Beginning the third week, parasite releases, at a female parasite-host ratio of 1:5, were initiated and continued until the week of 17 October 1983. The 1:5 parasite:host ratio was based on results obtained from earlier field releases (Morgan et al. 1981a,b).

At the second farm, 0.05% cyromazine was also included for two weeks in the diet of the caged layers, followed by the parasite releases at the same ratio and time interval as at the first farm.

The caged layers at the third installation were exposed to 0.05% cyromazine in their diet through the week of 17 October 1983. The fourth installation did not receive chemical treatments to suppress the fly populations and served as the untreated check.

The density of the adult *M. domestica* population was monitored weekly at each farm with a modified Scudder Grid (Murvosh & Thaggart 1966). The manure at all four farms was removed using a combination of mechanical equipment and manual labor.

RESULTS AND DISCUSSION

The combination of the cyromazine plus initiation of the parasite releases the third week resulted in a 90% reduction of the house fly pupal population during June at the first farm. The efficiency of the chemical was improved by the farmer's removing the manure prior to using the cyromazine and leaving a 4-inch pad of manure which provided protection for the indigenous parasites and predators. Although there was an increase in July, the augmentative release of the parasites continued to reduce the pupal population and reached a reduction of greater than 91% in September, 98% in October, and 99% in November. Parasitism increased from 25% during the pre-release sampling to 81% in October and 94% in November even though the parasite releases had been discontinued in October (Table 1).

Table 1. *Spalangia endius* and 1- and 2-day-old *Musca domestica* pupae populations as well as percentage parasitism of 1- and 2-day-old pupae at the first installation where the manure was allowed to cone naturally.[1,2]

1983	*S. endius* females	Parasitism (%)	1-,2-Day-old pupae	Reduction over pre-release (%)
Pre-release	--	25	1.0×10^6	--
June	6.3×10^4	79	1.1×10^5	89
July	2.9×10^4	75	2.0×10^4	98
Aug.	1.9×10^5	65	4.8×10^5	52
Sept.	1.5×10^5	35	9.0×10^4	91
Oct.	1.5×10^5	81	1.8×10^4	98
Nov.	1.3×10^3	94	1.2×10^4	99

[1] Release discontinued week of 17 October 1983.

[2] Cyromazine used for the first and second weeks, then parasite releases initiated the third week.

The pre-release parasite population was 100% *Muscidifurax raptor* G&S. Following initiation of the *S. endius* release in June and maintained through October, there was a dramatic increase in the *S. endius* population. The *M. raptor* population was reduced, and small populations of *S. cameroni* Perkins and *S. nigroaenea* Curtis became present. Based on the number of parasites recovered from wild fly pupae from June through December, the percentage was as follows: *S. endius* 46%; *S.cameroni* 1.8%; *S. nigroaenea* 0.2%; and *M. raptor* 52%.

During this same time interval, the adult fly population, of which many were flying in from other breeding areas, fluctuated from one to six flies/grid.

At the second farm, prior to incorporating the cyromazine into the diet of the caged layers, the owner cleaned the houses and pushed the accumulated manure along the outside of each house. However, the presence of cyromazine in the fresh poultry manure and the initiation of the *S. endius* releases gave only a 50% reduction in the pupal population during the first month. While there was only a slight decrease in the pupal population in July (10%), the parasite releases continued to reduce the pupal population, below the

Table 2. *Spalangia endius* **and 1- and 2-day-old** *Musca domestica* **pupae populations as well as percentage parasitism of 1- and 2-day-old pupae at the second installation where the manure was allowed to cone naturally.**[1,2]

1983	*S. endius* females	Parasitism (%)	1-,2-Day-old pupae	Reduction over pre-release (%)
Pre-release	--	30	7.4×10^5	--
June	5.9×10^4	80	3.7×10^5	50
July	6.1×10^4	14	3.1×10^5	60
Aug.	2.1×10^5	69	1.3×10^5	83
Sept.	1.4×10^5	56	4.8×10^3	99
Oct.	1.4×10^5	42	7.8×10^4	90
Nov.	1.2×10^3	78	7.7×10^4	90

[1] Release discontinued week of 17 October 1983.

[2] Cyromazine used the first and second week then parasite releases initiated the third week.

pre-release pupal population, to 83%, 99%, 90%, and 90% in August, September, October, and November, respectively.

Following initiation of the releases, parasitism rose to 80% in June, ranged from 14% to 69% from July through October, and increased to 78% in November (Table 2).

The pre-release parasite population was: *S. endius* 1%, *S. cameroni* 16%, *S. nigroaenea* 26%, and *M. raptor* 57%. Following initiation of the releases, the *S. endius* population increased to 39%, while the *S. cameroni, S. nigroaenea*, and *M. raptor* populations decreased to 15%, 6%, and 40%, respectively.

The adult house fly population during this same time interval ranged from one to fourteen flies/grid, again the result of migrating flies.

The pupal population at the third farm increased 1.6-fold in June and by October had increased an additional 8.1-fold. Following cessation of the treatments, the pupal population increased 63.3-fold the pre-release population (Table 3). The farm had a continuous problem with leaking water troughs and broken water pipes which kept the manure in a semi-liquid condition. The grid counts were taken in other areas of the poultry installation and ranged from two to eight/grid. Since parasites cannot locate fly

Table 3. 1- and 2-day-old *Musca domestica* L. pupae populations as well as percentage parasitism of 1- and 2-day-old pupae at the third installation where the manure was allowed to cone naturally.[1,2]

	Parasitism (%)	1-, 2-day-old pupae	Reduction over pre-release (%)
Pre-release	8	7.1×10^4	--
June	77	1.2×10^5	0
July	100	9.9×10^3	87
Aug.	35	1.3×10^5	0
Sept.	75	4.6×10^4	36
Oct.	13	9.8×10^5	0
Nov.	1	4.5×10^6	0

[1] Cyromazine used through 17 October 1983.

[2] Percentage parasitism (natural) obtained from wild fly pupae collected in dry areas only.

pupae in liquid manure, all pupae collected from the area of liquid manure were non-parasitized by indigenous parasites. Percentage parasitism of wild fly pupae collected in the dry areas peaked at 100% in July, then gradually decreased to 1% in November. Parasites that emerged from the pupae collected in the dry areas were *S. endius* (46%), *S. cameroni* (28%), and *M. raptor* (26%).

The fourth installation, which was used as a check, was free of insecticide treatments. The pupal population of 1.5×10^6 in June decreased to 60% in August but increased in September, and with the arrival of cool weather in October, increased two-fold the June population (Table 4).

The damp condition of the manure, resulting from leaking water troughs and inadequate protection from rainstorms, produced grid counts ranging from three to ten flies/grid from June through November.

Percentage parasitism of wild pupae collected from the dry areas near the damp manure was fairly high during July, August, and September but began decreasing with the arrival of cool weather in October and November. *Muscidifurax raptor* accounted for 65% of the parasitoid population, while *S. endius*, *S.cameroni,* and *S. nigroaenea* accounted for 19%, 3%, and 13% of the parasitoid population, respectively.

Table 4. 1- and 2-day-old *Musca domestica* L. pupae populations as well as percentage parasitism of 1- and 2-day-old pupae at the fourth installation where the manure was allowed to cone naturally.[1]

	Parasitism (%)	1-, 2-day-old pupae	Reduction over pre-release (%)
June	34	1.5×10^6	--
July	82	2.4×10^5	84
Aug.	63	6.1×10^5	60
Sept.	87	2.7×10^5	82
Oct.	51	1.5×10^6	0
Nov.	38	3.5×10^6	0

[1] Percentage parasitism (natural) of wild fly pupae collected in dry areas.

The temperatures for this area during the months of June, July, August, September, October, and November, according to the National Oceanic and Atmospheric Administration, were 24.7 ± 5.7, 27.5 ± 6.4, 27.5 ± 6.4, 23.9 ± 6.1, 20.5 ± 5.5, and 14.4 ± 7.8 degrees C, respectively.

DISCUSSION

The results of this study demonstrate that manure management is vital when using either chemical or biological control or combinations of both against populations of *M. domestica*. The method of manure management used by the owner of the first installation increased the effectiveness of the cyromazine and the parasite releases, which resulted in a 90% reduction of the *M. domestica* larval population during the first three weeks of the study.

The failure of the owner of the second farm to practice manure management reduced the initial effectiveness of both the cyromazine and the parasites. However, a 50% reduction in the *M. domestica* larval population was achieved during the first month. Continuation of the augmentative releases of the parasites through the following months further reduced the larval populations to a level equal to that obtained at the first farm. The parasite releases increased the *S. endius* population from 1.7% to 39%, while the populations of the other three species decreased.

The liquid condition of the manure at the third farm interfered with grid counts and prevented natural parasites and predators from locating the larvae and pupae, even though 100% parasitism was obtained from pupae collected in the dry areas near the installation in July. The sixty-three-fold increase in the larval population at the end of the treatment demonstrated that the liquid condition of the manure had reduced the effectiveness of the cyromazine. The house fly larvae that had been exposed to the cyromazine during the study (ca. 10 generations) would have had the opportunity to develop resistance as Bloomcamp et al. (1987) was able to demonstrate through laboratory studies.

The model developed by Morgan et al. (1981a,b) made it possible to determine the number of three-day-old third instar larvae that were present at the poultry installations. Monitoring the third instar larval population provided a reliable method of monitoring the effectiveness of the parasite releases. This knowledge of the third instar larval populations was used to determine the number of one-, two-, three-, and four-day-old pupae that were present at each poultry installation. Knowing the number of pupae available determined the number of female parasites that needed to be released to maintain a 1:5 parasite:host ratio.

Even after 126 generations in culture, the *S. endius* were competitive against other wild species of parasites. The *S. endius* populations increased from zero to 46%, and the *M. raptor* population dropped to 52%, while both populations of *S. cameroni* and *S. nigroaenea* increased.

The fourth farm, which served as a check, also had problems with manure management. Leaking water troughs and inadequate protection from rain storms kept the manure in a semi-liquid state and interfered with grid counts. Percentage parasitism of pupae collected from the dry areas, while high in July, August, and September, did not give an overall true indication of the percentage parasitism.

ACKNOWLEDGMENTS

The authors wish to express their appreciation to Gold Kist Feeds, Hillendale Farms, J. Crews, L. Crews, O. Crews, and G. McManus for allowing us to use their facilities to conduct the research study, and to G. Propp, J. Vaughan, D. Moore, and A. Benton for their assistance.

REFERENCES CITED

Bloomcamp, C. L., R. S. Patterson & P. G. Koehler. 1987. Cryomazine resistance in the house fly (Diptera: Muscidae). J. Econ. Entomol. 80:352-357.

Boucek, Z. 1963. A taxonomic study in *Spalangia* Latr. (Hymenoptera, Chalcidoidea). Acta Entomol. Musei Nationalis Pragae 35:429-512.

Hall, R. D. & M. C. Foeshe. 1980. Laboratory and field tests of CGA-72662 for control of the house fly and face fly in poultry, bovine or swine manure. J. Econ. Entomol. 73:564-569.

Morgan, P. B. 1981. Mass production of *Spalangia endius* Walker for augmentative and/or inoculative field releases, pp. 185-188. In Status of Biological Control of Filth Flies. Agric. Res., SEA, USDA, New Orleans, LA.

Morgan, P. B. & R. S. Patterson. 1977. Sustained releases of *Spalangia endius* to parasitize field populations of three species of filth breeding flies. J. Econ. Entomol. 70:450-452.

Morgan, P. B., R. S. Patterson & G. C. LaBrecque. 1976. Controlling house flies at a dairy installation by releasing a protelean parasitoid *Spalangia endius* Walker. J. Ga. Entomol. Soc. 11:39-43.

Morgan, P. B., R. S. Patterson, G. C. LaBrecque, D. E. Weidhaas, A. Benton & T. Whitfield. 1975a. Rearing and release of the house fly pupal parasite *Spalangia endius* Walker. Environ. Entomol. 4:609-611.

Morgan, P. B., R. S. Patterson, G. C. LaBrecque, D. E. Weidhaas & A. Benton. 1975b. Suppression of a field population of house flies with *Spalangia endius*. Science. 189:388-389.

Morgan, P. B., D. E. Weidhaas & R. S. Patterson. 1981a. Programmed releases of *Spalangia endius* and *Muscidifurax raptor* (Hymenoptera: Pteromalidae) against estimated populations of *Musca domestica* (Diptera: Muscidae). J. Med. Entomol. 18:158-166.

_____. 1981b. Host -Parasite Relationship: Augmentative releases of *Spalangia endius* Walker used in conjunction with population modeling to suppress field populations of *Musca domestica* L. (Hymenoptera: Pteromalidae and Diptera: Muscidae). J. Kans. Entomol. Soc. 54:496-504.

Mulla, M. S. & H. Axelrod. 1983a. Evaluation of the IGR Larvadex as a feed-through treatment for the control of pestiferous flies on poultry ranches. J. Econ. Entomol. 76:515-519.

_____. 1983b. Evaluation of Larvadex, a new IGR for the control of pestiferous flies on poultry ranches. J. Econ. Entomol. 76:520-524.

Murvosh, C. M. & C. W. Thaggard. 1966. Ecological studies of the house fly. Ann. Entomol. Soc. Am. 59:533-547.

Peck, O., Z. Boucek & A. Hoffer. 1964. Keys to the Chalcidoidea of Czechoslovakia (Insects: Hymenoptera). Mem. Entomol. Soc. Can. 34:170 pp.

Williams, R. E. & J. G. Berry. 1980. Evaluation of CGA-72662 as a topical spray and feed additive for controlling house flies breeding in chicken manure. Poultry Sci. 59:2206-2212.

FOOTNOTES

[1] A three-year pilot study financed by the USDA-ARS, "The Potential of an Integrated Pest Management Scheme to House Flies and Other Filth Breeding Flies at Poultry Farms with Special Emphasis on Use of the Parasitoid Wasp, *Spalangia endius* Walker.

[2] This paper reports the results of research only. Mention of a pesticide does not constitute a recommendation for use by the USDA nor does it imply registration under FIFRA as amended. Also, mention of a commercial or proprietary product does not constitute an endorsement by the USDA.

7. Native Biocontrol Agents as a Component of Integrated Pest Management for Confined Livestock

J. J. Petersen, D. W. Watson and B. M. Pawson

ABSTRACT

The management and operation of midwestern beef cattle confinements often results in substantial fly production. Further, these activities generally do not favor the build-up of naturally occurring pathogens and parasites. Little information is available on the pathogens of house flies or stable flies. The fungus, *Entomophthora muscae*, is the only pathogen observed to produce epizootics in house flies, and no epizootics have been observed in stable flies on midwestern confinements. Predators apparently have a significant impact on immature stages of these flies, but to date little is known about the species involved or the factors that influence their effectiveness as biological control agents. Pteromalid wasps parasitize the pupal stage of these flies and can be easily produced in large numbers. Because the numbers of these wasps begin at low levels and generally increase during the fly season, efforts have been made to enhance this build-up. Methods of establishing protected cohorts of house fly pupae in the field as host sources are discussed.

House flies (*Musca domestica* L.) and stable flies (*Stomoxys calcitrans* [L.]) have long been a problem associated with confined livestock. Considering the great reproductive potential of these flies, it is understandable that many natural mechanisms are involved in regulating their numbers. Legner & Brydon (1966) stated that it was astonishing that arguments still continue as to the relative merits of natural control in suppressing fly populations. Richards (1961) had previously estimated that if the sexes are produced in equal numbers, and a species lays 100 eggs, population stability results if 98% mortality occurs. A deviation of 0.5% in the mortality will result in a 25% increase or decrease in the population in the next generation. Thus, under normal conditions, an array of physical and biotic factors interact to limit the population growth of a species. When natural interactions are upset as with confined livestock (the accumulation of breeding sites and their continued disruption), natural controls fail to stabilize fly populations.

The resulting fly populations require alternate means of control such as habitat removal and the frequent use of insecticides. Adding to the problem, fly populations that previously may have been tolerable may no longer be acceptable because of urban encroachment. The combination of urban sprawl, reduced profit margin, and product contamination now require more sophisticated approaches in order to combat large numbers of adult flies. Understanding the naturally occurring biotic components, and the factors affecting them are necessary if the full potential of these components is to be realized.

Since the size, design and geographic location of the confinement greatly affect control strategies, much of the following discussion on natural biocontrol agents as a component of integrated fly management is limited to conditions found at midwestern beef cattle confinements.

Pathogens. The literature is surprisingly devoid of references to naturally occurring pathogens of house flies and stable flies, and no pathogens are currently available for use against these flies (Roberts & Strand 1977, Roberts et al. 1983). Apparently this area of research remains largely unexplored. Recent surveys in eastern Nebraska have isolated five species of bacteria that appear to be pathogenic to house flies and stable flies, but their impact on fly populations is unknown (unpublished information).

A strain of the fungus *Entomophthora muscae* (Cohn) Fresenius, isolated from house flies in Nebraska and morphologically similar to the New York strain (Kramer & Steinkraus 1987) has been observed to produce epizootics in adult house fly populations in the spring and particularly the fall. Similar seasonal activity was observed by Mullens et al. (1987) in a morphologically distinct strain of *E. muscae* in poultry facilities and dairy facilities (Mullens, this volume) in California. Infection levels of the Nebraska isolate have exceeded 70% in some host populations. Studies are presently underway to determine the environmental parameters that influence the spread of this disease and its impact on fly populations throughout the fly season.

Predators. Predators are acknowledged to be important elements of the natural control component of flies associated with confined livestock. A great diversity of arthropods utilize the immature stages of house flies and stable flies as food sources or compete with these flies for available resources. Considerable effort has been employed to define the impact of predatory arthropods on muscoid flies in habitats such as bovine droppings in pastures (Harris & Blume 1986, Doube 1986, Thomas & Morgan 1972), and poultry manure (Legner 1971, Pfeiffer & Axtell 1980, Axtell & Rutz 1986). However, few detailed studies on the impact of predators have been conducted on flies associated with confined beef and dairy cattle. In the most complete work to date, Smith et al. (1987) recovered forty-five species of arthropods from twelve families of Coleoptera, five of Diptera, two of Hemiptera, one of Dermaptera, and one mesostigmatid mite (Table 1), Macrochelids and staphylinids were the predacious arthropods most frequently collected. In earlier studies, Smith et al. (1985) determined egg-to-adult mortality for stable flies to be between 95 and 97%. Predation was found to be an important natural regulatory force that accounted for about 27% of the mortality. This mortality was associated principally with smaller coleopteran species, especially Staphylinidae. Similar mortality tables have been determined for house flies (Pers. comm. R. D. Hall, Univ. Missouri).

Predators and competitors are difficult to manipulate as biological control agents. Also, little is known about their population dynamics on open feedlots or the impact that livestock management practices have on these beneficial arthropods. Much research is needed before we have the knowledge to increase the effectiveness of these agents in controlling house flies and stable flies.

Parasites. Wasps of the family Pteromalidae are the primary parasites of flies associated with confined cattle, and most of the important species are solitary pupal parasites. Assuming that fly pupae represents 1 - 5% of the total house fly and stable fly eggs deposited (Smith et al. 1985), and that the previously stated assumption of Richards (1961) is true, then pteromalids must exert a greater natural control effect per individual than an agent destroying any previous stage (Legner & Brydon 1966). Though pteromalid wasps have been extensively studied for a number of years, most of the research has been limited to laboratory observations or attempts to control flies through sequential releases of laboratory reared wasps. In the last five to ten years additional important data have been gathered on the biology and ecology of pteromalids in nature.

Parasitism by pteromalids of filth flies associated with livestock is usually low, and mean seasonal parasitism is usually below 15% (Rueda & Axtell 1985, Petersen & Meyer 1983a). Studies by Guzman & Petersen (1986) and Petersen & Meyer (1983b) suggest

Table 1. Competitor and predaceous arthropods associated with immature stages of stable flies in Missouri (after Smith et al. 1987).

Association	Order	Family	No.species
Competitor	Dermaptera	Labiidae	1
	Hemiptera	Cydnidae	1
		Anthocoridae	1
	Coleoptera	Scarabaeidae	2
		Elateridae	2
		Rhizophagidae	1
		Cucujidae	1
		Mycetophagidae	1
		Tenebrionidae	2
		Anthicidae	1
		Curculionidae	1
	Diptera	Syrphidae	1
		Otitidae	1
		Sepsidae	1
		Muscidae	8
		Sarcophagidae	1
Predators	Coleoptera	Staphylinidae	8
		Histeridae	4
		Carabidae	5
		Hydrophilidae	1
	Mesostigmata	Macrochelidae	1

Table 2. **Percent parasitism of house fly and stable fly pupae by the principal species of pteromalid wasps at feedlots in eastern Nebraska (after Petersen & Meyer 1983)**

Species	May	June	July	Aug	Sept	Oct
M. zaraptor	0	2.7	2.9	4.5	5.3	9.7
S. cameroni	0	1.0	0.4	1.8	1.8	1.2
S. nigroaenea	0	1.1	1.2	3.7	4.2	4.6
Mean	0	4.8	4.5	10.0	11.3	15.5

that overwintering survival of pteromalid wasps is low on midwestern U. S. cattle confinements. Reasons for the low survival include the very cold weather and the removal or destruction of overwintering habitats. Thus, parasite activity in the early part of the fly season is very low, usually below 1-5% (Table 2). Under suitable conditions, parasite populations increase throughout the fly season, often parasitizing a substantial portion of the host population by October (Figure 1). Our recent unpublished observations suggest that midsummer temperatures may have a deleterious effect on parasite survival. If this is the case, then when house flies and stable flies are at peak production (late spring and early fall), parasite populations are often low because of their protracted rate of development. To compensate for overwintering and possible midseason reductions, and to increase populations to levels where they have a significant impact on host populations, these parasites must be manipulated either by providing safe, stable developmental sites or by augmenting with laboratory reared wasps.

Generally, parasite releases through distribution of parasitized host pupae have been the approach of choice. However, this method usually has not been effective for a variety of reasons. These include the use of species poorly adapted to the particular host, host habitat or local climatic conditions; poor quality parasites (or contaminant species) resulting from too many generations in culture or from poor culturing techniques; or releases made under less than favorable climatic conditions. Perhaps a better approach

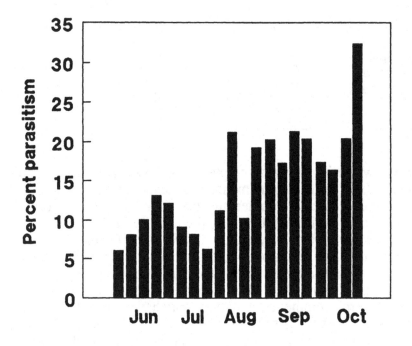

Figure 1. Mean parasitism by pteromalid wasps of house fly pupae collected from beef cattle feedlots and dairies in eastern Nebraska over the fly season (after Petersen & Meyer 1983).

may be to build up wasp numbers in the field by providing stable host sources. Most likely, parasites reared under protected field conditions will be better adapted to the specific hosts, host habitats and climatic conditions. As a result, considerable research effort has been expended at the Midwest Livestock Insects Research Laboratory to develop techniques for the field rearing of pteromalid wasps.

Observations on the overwintering biology of these parasites in eastern Nebraska suggested that at least some species were utilizing late season freeze-killed (f-k) fly pupae as hosts (Petersen & Meyer 1983b). Further, these observations suggested the possible introduction of f-k house fly pupae for the early season build-up of naturally occurring parasites. An introduced host source of f-k pupae would remain at a stage suitable for utilization for a longer period than would be the case with live pupae. Also, f-k hosts would not add to the existing adult fly population. In laboratory studies, *Muscidifurax zaraptor* Kogan and Legner readily utilized f-k house fly pupae, and progeny production was equal to that when using live house fly pupae (Petersen & Matthews 1984). Under laboratory conditions, *M. zaraptor* discriminated between live and f-k house fly pupae at low parasite-to-host ratios (1:40) and preferred live hosts; however, discrimination was not evident at high parasite-to-host ratios (1:5) (Petersen et al. 1986). Under field conditions when f-k house fly pupae were placed adjacent to emerging *M. zaraptor, Spalangia cameroni* Perkins and *Urolepis rufipes* (Ashmead), *S. cameroni* and *U. rufipes* did not use the f-k pupae as hosts, but *M. zaraptor* readily oviposited on and produced progeny on the introduced hosts (Petersen 1986). Additional studies showed that *M. zaraptor* would disperse and search for other hosts even when large numbers of f-k hosts were available. *M. zaraptor* readily moved out and parasitized f-k hosts placed six m from the release site in the four cardinal directions, and levels of parasitism were similar but lower than those for hosts placed adjacent to the emerging parasites (Petersen & Pawson 1988).

A study was made during the summer of 1987 in an attempt to increase native parasite populations by establishing artificial host sources. Two methods were studied. The first method was to establish protected sources of live and f-k house fly pupae and add additional hosts periodically to permit a natural build-up of the naturally occurring parasite populations. The second method was to make a single release of laboratory-reared, native *M. zaraptor* adjacent to f-k house fly pupae and to increase the numbers of f-k pupae over time in an attempt to increase parasite numbers.

To determine if naturally occurring parasites could be increased, cohorts of 2000 f-k (frozen when forty-eight to seventy-two h old) and 2000 live (two to twenty-four h old) house fly pupae were placed separately in fiberglass window screen (eighteen mesh) bags. The paired cohorts of fly pupae were placed at four stations on each of two dairies. The bags were placed in the habitat, lightly covered with habitat material, and covered by an open bottomed cage (20 x 30 x 30 cm) constructed of 0.6 cm wire screen supporting a plywood top. This shelter protected the pupae from adverse climatic conditions and reduced the chance of damage by rodents. Additional cohorts of 2000 live and f-k host pupae were added to each station every two weeks over a sixteen week period. Samples of 100 pupae were removed each week from each cohort to measure the level of parasitism.

Many of the cohorts were lost or destroyed as a result of changes in dairy management practices during the study. These losses greatly disrupted the continuum of exposed cohorts necessary to maximize increases in the parasite populations. However, parasite emergence from f-k hosts increased from 1 to 44% over the sixteen week test period (Figure 2). *Muscidifurax raptor* Girault and Sanders and *M. zaraptor* comprised 99% of the parasite guild using the f-k house fly pupae as hosts. *Pachycrepoideus vindemmiae* (Rondani) was the only other species recovered from the f-k hosts. Surprisingly, parasite emergence from live hosts increased only to about 23% over the same period. Again the *Muscidifurax* spp. made up the majority (97%) of the parasite guild using the introduced live house fly pupae as hosts. The balance was made up of *P. vindemmiae* and *Spalangia nigroaenea* Curtis. The results suggest that creating stable host sources for native parasites has the potential to increase parasite populations substantially. However, the rate of buildup of parasite numbers appears to be insufficient to impact fly populations during the early and middle of the fly developmental season.

In the second part of the study, attempts were made to mass produce parasites under field conditions. To accomplish this, four rearing stations were established at each of two facilities, a beef cattle confinement and a dairy. Stations were similar to those employed for the native parasitism studies. On 7 May, cohorts of 2000, 4000, 6000 or 8000 f-k house fly pupae (frozen when forty-eight to seventy-two h old) were placed at the four stations on both confinements. At each station, a single cohort of 2700 house fly pupae parasitized by *M. zaraptor* (parasitized in the laboratory), was placed adjacent to each cohort of f-k hosts. Emerging female

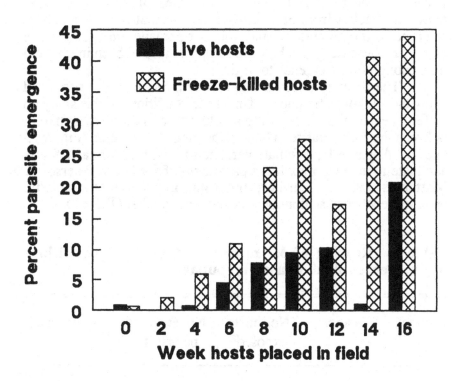

Figure 2. Mean parasitism by native pteromalid wasps of living and freeze-killed house fly pupae placed at four locations on each of two dairies over a sixteen week period.

parasites resulted in parasite-host ratios of about 1:2, 1:4, 1:6 and 1:8. Additional cohorts of 2000, 4000, 6000 and 8000 f-k hosts were placed at the corresponding stations at both locations after two and four weeks. Then, six, eight and ten weeks after the start of the study, 20,000 f-k hosts were placed at each station at both locations. An additional 20,000 f-k hosts were placed at the four stations at the dairy only on the twelfth week. Samples of 100 pupae were collected weekly from each host cohort present at each location, returned to the laboratory, and held for parasite emergence or to count parasite emergence holes in samples that had been in the field long enough for emergence to begin.

The results indicated that *M. zaraptor* readily reproduced using the f-k house fly pupae. Under the conditions of this study no differences in the extent of parasite emergence were apparent between the four parasite to host ratios for either location (Figures 3 and 4). Apparently, a female parasite-to-host ratio of ca. 1:8 was adequate to achieve maximum parasitism of f-k hosts (Petersen & Matthews 1984). Further, the total mean percent parasite emergencebetween stations at a location was similar (Table 3).

Table 3. Rearing of *M. zaraptor* under field conditions using freeze-killed house fly pupae.

Location	Station	No. hosts exposed[a]	No. emerged parasites	% emergence
Feedlot	1	66,000	25,320	38.4
	2	32,000[b]	12,000	37.5
	3	78,000	31,000	39.7
	4	84,000	32,900	39.2
Dairy	1	86,000	39,060	45.4
	2	92,000	40,720	44.2
	3	98,000	47,600	48.6
	4	104,000	53,240	51.1

[a] 2000, 4000, 6000 and 8000 host pupae were placed in stations 1, 2, 3 and 4, respectively, for the first three releases, and 20,000 host pupae in each station in subsequent releases.

[b] Station destroyed after fourth release.

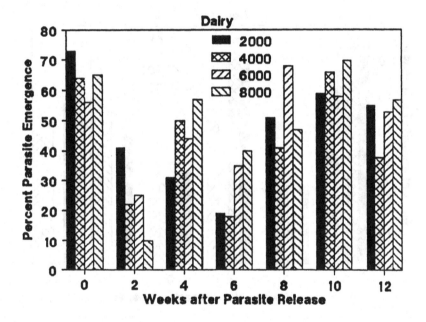

Figure 3. Parasitism of freeze-killed house fly pupae placed at a dairy seven times over a twelve week period following a single release of *M. zaraptor* at week zero. Numbers of fly pupae ranged from 2000 to 8000 in four sites for the first three releases; all sites received 20,000 pupae for all subsequent releases.

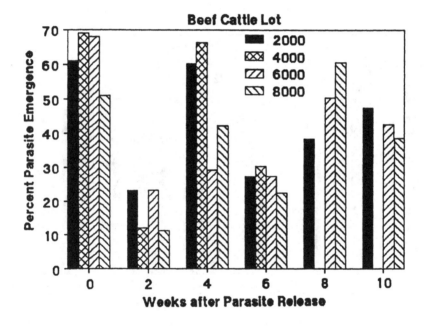

Figure 4. Parasitism of freeze-killed house fly pupae placed at a beef cattle feedlot six times over a ten week period following a single release of *M. zaraptor* at week zero. Numbers of fly pupae ranged from 2000 to 8000 in four sites for the first three releases; all sites received 20,000 pupae for all subsequent releases.

Reduced parasitism in cohorts placed out weeks two and six apparently resulted from low numbers of *M. zaraptor* in the stations because progeny from previously parasitized f-k pupae had not completed development. Observations during the study suggested that desiccation, especially in the more exposed areas during mid summer, can greatly reduce parasite survival. Though these studies suggest that considerable potential exists in increasing the effectiveness of native pteromalids, especially *M. zaraptor*, through habitat modification and introduced hosts, considerable refinement of the procedures will be necessary to make this method of fly control cost effective and practical.

Natural control is an important and necessary component of any integrated pest management program. This component is often ignored when considering control approaches. Perhaps this results from our lack of understanding of the factors involved and how livestock management and fly control practices impact native biocontrol agents. A greater research effort is needed in this area to develop a better understanding of this relationship before the effectiveness of native agents can be maximized.

REFERENCES CITED

Axtell, R. C. & D. A. Rutz. 1986. Role of parasites and predators as biological fly control agents in poultry production facilities. Misc. Publ. Entomol. Soc. Am. 61: 88-100.

Doube, M. 1986. Biological control of the buffalo fly in Australia: The potential of the southern Africa dung fauna. Misc. Publ. Entomol. Soc. Am. 61: 16-34.

Guzman, D. R. & J. J. Petersen. 1986. Overwintering of filth fly parasites (Hymenoptera: Pteromalidae) in open silage in eastern Nebraska. Environ. Entomol. 15: 1296-1300.

Harris, R. L. & R. R. Blume. 1986. Beneficial arthropods inhabiting bovine droppings in the United States. Misc. Publ. Entomol. Soc. Am. 61: 10-15.

Kramer, J. P. & D. C. Steinkraus. 1987. Experimental induction of the mycosis caused by *Entomophthora muscae* in a population of house flies (*Musca domestica*) within a poultry building. J. New York Entomol. Soc. 95: 114-117.

Legner, E. F. 1971. Some effects of the ambient arthropod complex on the density and potential parasitization of muscoid Diptera in poultry wastes. J. Econ. Entomol. 64:111-115.

92

Legner, E. F. & H. W. Brydon. 1966. Suppression of dung-inhabiting fly populations by pupal parasites. Ann. Entomol. Soc. Am. 59: 638-651.

Mullens, B. A., J. L. Rodriguez & J. A. Meyer. 1987. An epizootiological study of *Entomophthora muscae* in muscoid fly populations on southern California poultry facilities, with emphasis on *Musca domestica*. Hilgardia 55: 1-41.

Petersen, J. J. 1986. Augmentation of early season releases of filth fly (Diptera: Muscidae) parasites Hymenoptera: Pteromalidae) with freeze-killed hosts. Environ. Entomol. 15: 590-593.

Petersen, J. J. & J. R. Matthews. 1984. Effects of freezing of host pupae on the production of progeny by the filth fly parasite *Muscidifurax zaraptor* (Hymenoptera: Pteromalidae). J. Kansas Entomol. Soc. 57: 387-393.

Petersen, J. J. & J. A. Meyer. 1983a. Host preference and seasonal distribution of pteromalid parasites (Hymenoptera: Pteromalidae) of stable flies and house flies (Diptera: Muscidae) associated with confined livestock in eastern Nebraska. Environ. Entomol. 12: 567-571.

_____. 1983b. Observations on the overwintering pupal parasites of filth flies associated with open silage in eastern Nebraska. Southwest. Entomol. 8: 219-225.

Petersen, J. J. & B. M. Pawson. 1988. Early season dispersal of *Muscidifurax zaraptor* (Hymenoptera: Pteromalidae) utilizing freeze-killed housefly pupae as hosts. Med. Vet. Entomol. 2: 137-140.

Petersen, J. J., B. M. Pawson & D. R. Guzman. 1986. Discrimination by the pupal parasite *Muscidifurax zaraptor* for live and freeze-killed house fly pupae. J. Entomol. Sci. 21: 52-55.

Pfeiffer, D. G. & R. C. Axtell. 1980. Coleoptera of poultry manure on caged-layer houses in North Carolina. Environ. Entomol. 9: 21-28.

Richards, 0. U. 1961. The theoretical and practical study of natural insect populations. Ann. Rev. Entomol. 6: 147-162.

Roberts D. W. & M. A. Strand. 1977. Pathogens of medically important arthropods. Bull. Wld. Hlth. Org. SS (Suppl. 1), 419 pp.

Roberts, D. W., R. A Daoust & S. P. Wraight. 1983. Bibliography on pathogens of medically important arthropods: 1981. Wld. Hlth. Org. VBC/83.1, 324 pp.

Rueda, L. M. & R. C. Axtell. 1985. Comparison of hymenopterous parasites of house fly, *Musca domestica* (Diptera: Muscidae), pupae in different livestock and poultry production systems. Environ. Entomol. 14: 217-222.

Smith, J. P., R. D. Hall & G. D. Thomas. 1985. Field studies on mortality of the immature stages of the stable fly (Diptera: Muscidae). Environ. Entomol. 14: 881-890.

_____. 1987. Arthropod predators and competitors of the stable fly *Stomoxys calcitrans* (L.) (Diptera: Muscidae) in central Missouri. J. Kansas Entomol. Soc. 60: 562-567.

Thomas, G. D. & C. E. Morgan. 1972. Field-mortality studies of the immature stages of the horn fly in Missouri. Environ. Entomol. 1: 453-459.

8. Puparial Factors in Host Location by *Spalangia endius* (Walker) (Hymenoptera: Pteromalidae)

Martin J. Rice

ABSTRACT

The literature contains contradictions regarding the importance of puparial factors in host location by parasitoids of muscoid flies. Several olfactometers were designed to enable this matter to be resolved; the most effective design found was the "sandwich choicer," built from the tops of tissue culture plates and small petri dishes. Details are given of the sandwich choicer construction and operation, and of a simple environmental chamber suited for the olfactory testing of parasitoids.

Using a Rate of Selective Attraction (ROSA) index, the parasitoids were found to be selectively attracted by volatiles from whole puparia of *Musca domestica* L. and *Phaenicia (Lucilia) pallescens* (Shannon). ROSA indexes found were: whole puparial volatiles = 5.2%/h; larval medium volatiles = 3.7%/h; 500 ppm ammonia vapor = 2.9%/h; water vapor = 2.6%/h; and the volatiles from a solvent extract of puparial cuticle = 1.7%/h. It was concluded that puparial volatiles are likely to play a role in the proximate phase of host location.

The very regular rate of attraction of members of the parasitoid population is considered to reflect the operation of central nervous system "attention switching" circuitry.

INTRODUCTION

Johnson and Tiegs (1922) and Miller (1927) proposed biological control of sheep blowflies using puparial parasitoids. The potential of pteromalid microhymenopteran puparial parasitoids as control agents for pest populations of the house fly *Musca domestica* L. has been considered by Legner and Detrick (1972). However, a

95

single massive release of such puparial parasitoids failed to control house flies on Danish farms (Mourier 1972). Sustained releases of a laboratory mass-reared field strain of *Spalangia endius* Walker have, however, been found to be capable of eradicating localized populations of *M. domestica* at commercial poultry installations in the U.S.A. (Morgan et al. 1975a,b). In addition, this species has been successfully used to substantially reduce the numbers of *M. domestica, Stomoxys calcitrans* L. and *Physiphora aeneae* (F.) at dairy and beef installations (Morgan et al. 1976a,b). Such results have gone some way towards confirming the theory that parasitoids are likely to be a successful control agent if adequate numbers are sustained (Knipling and Brydon 1966). Thus it has become of some importance to understand more of the behavioral and environmental factors influencing the number of *S. endius* available to oviposit in *M. domestica* puparia, so that pest management decisions can be made on the minimum number of parasitoids needed for release, in a variety of control situations. If adequate numbers of parasitoids are released, fly density may be kept so low that little or no recourse may be necessary to chemical control, with its associated problems of toxicity, residues and resistance.

Weidhaas et al. (1977) originated a model that permits the study of the numerical relationships between *S. endius* and *M. domestica* under a variety of conditions. Among the specific objectives of their numerical model was the obtaining of a better understanding of the parasitoid/host interaction. In order for the model to function certain variables, such as patterns of egg-laying, had to be treated as constants. The authors observed that as further data became available on such variables, the simulation could be adapted to incorporate them. Thus the need for a technique to enable a closer study of the factors regulating the host locating behavior of *S. endius* is apparent. This point is emphasized by the findings of Morgan et al. (1981) who show that, under field conditions, the interrelationship of different pteromalid parasitoids and the puparia of *M. domestica* depend on mechanical and environmental factors. This finding was further highlighted by the discovery that the strain of *S. endius* previously so effective in a variety of release situations was not effective in controlling house fly and stable fly populations under other circumstances (Petersen et al. 1983). The situation where we have a "good" parasitoid credited with well documented successes, plus examples of its failure, has been the stimulus for the current work. This paper describes a simple laboratory test apparatus that has proved suitable for investigation of the behavior of small populations of *S. endius,* under a variety of environmental, physical and host stimulus situations. The apparatus is tendered as a contribution towards the design of a useful apparatus for the elucidation of the factors that make a particular parasitoid species or strain successful in some circumstance and not in others. Microhymenopteran parasitoids are difficult subjects for laboratory choice experiments because of their diminutive size, rapid and erratic locomotion, pronounced escape behavior and aggregative interactions. It may be helpful in screening parasitoids for field release if more of their choice behavior could be discovered by means of simple laboratory tests such as the preliminary ones demonstrated here.

Several workers have concluded that parasitoids are not attracted to puparia (e.g. Wylie 1958; Murphy 1982; and Stafford et al. 1984). They concluded that parasitoids locate the source of odors emanating from fly breeding sites and then encounter puparia by randomly burrowing into the media. However, Murphy also adduced evidence of a statistically significant attraction of *Spalangia cameroni* to house fly puparia. It therefore seemed worthwhile to attempt to address the question of pupal attractiveness for *S. endius,* as a practical test of the olfactometers in the present work.

MATERIALS AND METHODS

The High Springs, Florida, strain of *Spalangia endius* Walker and a laboratory strain of *Musca domestica* L. were cultured by the methods of Morgan et al. (1978). Wild caught *Phaenicia pallescens* (Shannon) was cultured in chopped bovine liver, over sawdust. The parasitoids were used zero to five days after eclosion. They were attracted to light and then captured and held in one cm diameter polystyrene tubes with press-snap, polyethylene caps pierced by ten 0.1 mm ventilation holes. When house fly or blow fly puparia were used as a lure, these were taken one to two days after pupariation and repeatedly washed in several changes of deionized water until all traces of the rearing medium were removed; they were then dried. A number of olfactory choice chambers were designed. One of these was based on a 10 x 10 x 12 inch plexiglass box, having two stainless steel mesh-covered, 2 inch diameter ventilation ports. Several hundred *S. endius* males and females were released in this box. Lures were presented in paper coffee cups, the insides of which were coated with Tangle-Trap®

adhesive. The tops of the cups were sealed with a polystyrene petri dish top; a central 2 mm diameter hole allowed egress of odors and ingress of any parasitoids attracted. The parasitoids entering the "cup trap" were soon entangled in the adhesive, making it easy to count their numbers after a suitable period of exposure.

Another of the olfactory choice chambers designed was the "eight choicer." This was composed of a series of eight, 2.5 inch long, 1 cm in diameter, polystyrene tubes, evenly spaced-out and heat welded onto the surface of a 91 mm diameter polystyrene petri dish top (Falcon 1029). Holes of 7 mm diameter were pierced through the petri dish to allow access into the tubes, the tops of the tubes being sealed with "snap-caps" as described previously. Approximately halfway up their length the tubes were loosely blocked by balls of non-absorbent cotton wool. Lures were placed above the cotton wool, where they were not visible to the parasitoids, which were introduced into a paper ice cream container fitted under the petri dish. The parasitoids moved around the paper container and entered the basal holes of the projecting tubes. If the lure was an attractant, this was signified by an increase in the numbers of parasitoids aggregating in the tube. In addition the insects made efforts to burrow past the cotton wool plugs into the lure-containing section of the tubes. A great advantage of the eight choicer is the built-in experimental replication achieved by using the same lure in four of the choice tubes, each alternating with a control tube. This type of replication is especially useful in situations where it is suspected that a slight lateral thermal or light intensity bias may be present.

The third design of olfactory choice chamber, described here as the "sandwich choicer," was the one most extensively used in the present study (Figures 1 and 2). This device was constructed from two polystyrene tissue culture plate lids (Falcon 3047), joined together by adhesive tape at the ends to form a 127 x 85 x 25 mm flat chamber. This had a 15 mm diameter hole pierced in the center of the top, which was generally sealed by an "O" gauge rubber bung. Two 57 mm diameter polystyrene petri dishes (Falcon 1007) were heat sealed to the underneath, and then five 2 mm diameter holes were pierced through each petri dish lid into the underside of the starting chamber. After sealing and piercing, the plastic was thoroughly washed and dried to remove unwanted odors. Thus the interiors of the petri dishes were in communication with the interior of the flat (starting) chamber (Figure 1). The inside surfaces of both the top and the bottom of these petri dishes were coated with an approximately 1 mm thick layer of Tangle-Trap adhesive. The

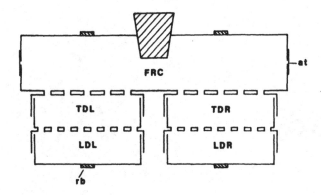

**Figure 1: Diagrammatic cross-section of the
sandwich choicer showing "three-storied"
construction.** Flat (starting) chamber at the top
(FRC), made of two tissue culture plate covers, taped
together at their ends (at), with a central bung hole in
the top for introduction of parasitoids. Right and left
trap dishes (TD) in the middle, connected to FRC by
five holes and to LD by 12 holes, internal surfaces
coated in adhesive. Right and left lure dishes (LD) at
the bottom. Rubber bands (RB) hold the components
together; removal of the bands enables the trap and
lure dishes to be opened for counting and servicing.

Figure 2: Photograph of a sandwich choicer opened up to reveal the components. Note the two tissue culture plate lids at the right of the figure, one with central bung hole and the other fused to trap dish tops. The bottoms of these dishes occupy the middle of the figure; trapped *S. endius* can be seen stuck in the Tangle-Trap adhesive in both halves of the trap dishes. The tops of the lure dishes are fused under the bottoms of the trap dishes; the holes connecting the two can be seen. The bottoms of the lure dishes occupy the left part of the figure: the bottom one contains clean puparia of *M. domestica,* the top one is an empty control dish. Note the distribution of the trapped parasitoids, almost exclusively above the puparia.

undersurfaces of the petri dishes were heat sealed to the tops of similar 57 mm dishes and twelve 2 mm diameter holes were pierced through, so that the insides of the top and bottom petri dishes communicated. This "sandwich" design permits lures to be presented in the lower petri dishes, their odors diffusing up through the middle adhesive-coated dishes to enter the top starting (flat) chamber.

Sandwich choicers are like three-storey buildings: the ground floor containing the lures, the first floor the sticky traps and the second floor the exposed population of parasitoids. The parasitoids walk around the top (starting) chamber at a rate dependent on their degree of dehydration and on the ambient temperature, up to approximately 10 mm per second. During this activity they periodically pass the two sets of five openings that connect to the two trap chambers below. If they are attracted to enter a trap chamber, they then generally become entangled in the adhesive. A small proportion of the parasitoids proved capable of negotiating the adhesive trap and enter the lowest, lure-containing dish. Each hour the numbers of parasitoids remaining in the top chamber, stuck in the middle dishes, or free in the lure dishes are counted (Figure 2). From this the percentages of the population attracted by the lures on the two sides of the choicer are calculated. The rate at which the insects are attracted is found by plotting their entry rates for twelve to twenty-four hours. Thus the attractiveness of any lure can be expressed either as the percentage of the population choosing that side, that is, the final total, once the experiment is concluded; or as the percentage of the available population choosing that side per hour, that is, their absolute rate of entry. By comparing rates of entry between two sides, a preference index can be derived; if this is done in relation to the numbers from which the choosers came, a rate of selective attraction (ROSA) is obtained.

The olfactory choice chambers were placed in constant temperature cabinets for the duration of the tests. The most effective type of cabinet tested was a modified laboratory oven, it had a temperature of $32\pm 2°C$ and had an unusual lighting system, with a diffused, 25 watt light source situated below the test apparatus. The light was blocked from shining directly up into the choice chambers by a piece of black cardboard (Figure 3). This design established a light gradient that was strongest from the sides (approximately 1-2 footcandles at the test chamber) and weakest from directly above (approximately 0.2 footcandles at the test chamber). Directly above the diffuser dome, the center of the black cardboard was 2°C warmer

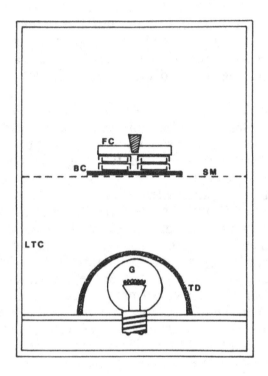

Figure 3: Diagrammatic cross-section of constant light and temperature cabinet (LTC) with 25 watt globe (G) under a translucent dome (TD) to diffuse the light; a sandwich choicer (FC), is supported by black cardboard (BC), at the center of a perforated shelf (SM).

than the rest of the cabinet; this may have caused a convection current in the sandwich choicer, carrying odors up towards the starting chamber. Although every effort was made to make this arrangement symmetrical, the test chambers were rotated 180° each hour to compensate for any slight irregularities in illumination and temperature.

In calculating the comparative effectiveness of attractants, from the results obtained with the flat choicer, the following formula was used:

$$ROSA^{A/B} = (A - B) \times 100 / (A + B + C) \times T$$

$ROSA^{A/B}$: rate of selective attraction of A over B;
A: the number of insects making the higher choice;
B: the number of insects making the lower choice;
C: the residual number not making a choice;
T: the duration in hours of the test.

This basically gives an index of the percentage of the available population that chose A over B per hour. The ROSA index could also be used to quantify the relative effectiveness of repellents, if A is made the lower number attracted and B the higher number. Several indexes of preference/attraction and non-preference/ repellence are commonly used to process choice experiment data. The main advantage of the new ROSA index is that it permits direct comparisons between different lures, environments and even different species. It is an index of the rate of depletion of an available population. Unfortunately, choice experiments are usually run for variable amounts of time and then read, possibly wasting valuable comparative data on rates of attraction. It is hoped that more behaviorists will have recourse to ROSA indexes, to facilitate comparisons between different species, stimuli and environments. The prolonged hours of observation needed could be divided between several people.

The solvent extract of 2-day *M. domestica* puparia was made by rinsing 100 clean puparia with 10 ml of a mixture of equal parts of "Analar" grade ethanol, petroleum ether and acetone; the solvents were evaporated at room temperature for twenty-four hours and the residue taken up on filter paper.

RESULTS

All the results given here are from groups of forty to seventy male and female *S. endius*, placed in the top (starting) chamber of sandwich choicers, which were then kept in the specialized environment cabinet described above. Hourly counting of the numbers of parasitoids caught in each of the two trap chambers revealed a remarkably steady rate of ingress of the insects. In a no-choice situation, about 2% per hour of the available population descended towards blank (dry, empty) chambers; about 4% per hour towards deionized water containing chambers; and about 6% per hour towards puparia (sixty to eighty super-clean and dry).

Thus there is always a steady rate of entry by parasites into non-lure-containing "dry" chambers. When deionized water is present their rate of entry is increased and when cleaned puparia of *M. domestica* are present their rate is even higher. After eighteen hours of exposure, these rates of entry typically resulted in about 30% of the population being trapped by no lure, 60% by water vapor and over 90% by the puparial odor. Plotted on linear axes, these results yielded straight lines.

In ten, prolonged choice experiments, using the ROSA index, the following rates of specific attraction were obtained:

		ROSA pref.for A/hr
1.	A: dry; B: dry:	0.08%
2.	A: *Musca* puparia; B: dry:	5.2%
3.	A: water; B: dry:	2.6%
4.	A: puparial extract ; B: dry:	1.7%
5.	A: puparia + water; B: water:	2.0%
6.	A: 2nd larval medium; B: dry:	3.7%
7.	A: 2nd larval med.; B: water:	0.8%
8.	A: 500 ppm NH_4OH; B: water:	2.9%
9.	A: 250 ppm NH_4OH; B: water:	2.2%
10.	A: *Phaenicia* pup.; B: *Musca* pup.:	0.1%

Parasitoids were found to have their highest specific rate of attraction to puparia, there being little to choose between *Phaenicia pallescens* and *Musca domestica* puparia. Larval medium from around second instar *M. domestica* larvae was next in attractiveness, followed by dilute ammonia and water (both are present in the second larval medium). Cuticular extracts of *M. domestica* puparia were also significantly attractive, though not as much as intact puparia. Puparial cuticular volatiles, excreted ammonia and transpired water vapor all may contribute to the high attraction to whole puparia. There are likely to be other volatiles involved too. The high attractiveness of dilute ammonia plus water vapors, even when offered against water as the "B" choice (expts 8 and 9), is surprising. Further work is needed to replicate the experiments, provide information on the choice "puparia versus ammonia," and investigate other puparial derived volatiles.

CONCLUSIONS

The ability of the "sandwich choicer" to resolve differences in attractiveness of various lures for *S. endius* has been shown. Some of the advantages of this type of olfactory choice device have been outlined in the Materials and Methods section above. Kyi (in an unpublished report, Entomology Department, University of Queensland 1987) compared the sandwich choicer with a complex olfactometer (Vett et al. 1983) and concluded that the sandwich choicer had advantages for investigating the behavior of populations of small parasitoids like *Spalangia*.

The slow rate of entry (usually about two to three parasitoids per hour) is remarkable but it is worth noting that the parasitoids made their choices at a distance of more than 30 mm, with no air stream and without visual cues to the presence of the puparia. They overrode their normal escape reactions and also their avoidance reactions to sticky media in order to enter the trap dishes to be caught.

Jacobi (1939) seems to have been the first to suggest that pupal volatiles attract parasitoids and recently Hogsette & Butler (1981) used puparia to attract parasitoids in field traps. *S. endius* will enter the sandwich choicer traps at a steady rate, by exploration, even in the absence of puparia. However, the presence of attractants induced increased rates of entry, with puparia yielding the highest ROSA indexes identified in this preliminary study.

A possible disadvantage of the system used here is the danger of spurious results, caused by the hourly removal of the sandwich choicers to count the *S. endius* trapped. However, the parasitoids enter the trap dishes at a remarkably steady rate, whether the choicers were examined frequently or infrequently. This suggests that individuals in the population reach the point of decision to move down into the lower chamber at widely varying times. In bygone years the concept of varied "motivation" might have been invoked. However, the most productive way to consider this phenomenon may be to recognize it as the outcome of a regular, internal scanning of different possible modes of behavior by the central nervous system. It would thus appear that the internal "attention" deciding neuronal circuitry independently switches individual insects into a downward seeking behavioral mode, leading to trap entry, so giving the remarkable uniformity in percentage trapped with time. Alternative behaviors able to be switched on include: escape, walking, flying, cleaning, social interactions, and remaining still. An attention switch system has recently been incorporated into a model of behavior, to replace concepts of "motivation," "action specific energy," and "central excitatory state" (Rice 1989). Differences in attention switching may help explain the marked individual variability of behavioral responses in an apparently uniform species population. This will only be confirmed once the neural and genetical bases for attention switching have been identified and their inter- and intra-specific variations determined.

The orientation behavior of parasitoids is a well-worked area (cf. Vinson 1976) and there have been several useful designs for olfactometers, for example, those of Varley and Edwards (1953) and Vett et al. (1983). Dethier (1947) described a number of olfactometers and choice devices for use with a range of insects. These are useful for studying the behavior of an individual insect, or a small number, being of particular use in behavioral analysis of insect orientation mechanisms. However, the work described here has been directed towards the development of a routine bioassay system, suited to comparative studies of small populations of parasitoids. This approach necessitated the evolution of an olfactory choicer suitable for screening mass responses of parasitoids, to a range of potential attractants, and operable in a range of environmental conditions. The aim has been to have a laboratory test that will usefully predict differences that are relevant to the behavior of populations of particular strains of species that are being considered for field biocontrol use. The sandwich choicer is the best such device tested in the current work, though there is no doubt that it could be improved. It has the advantage that various materials can be packed into the top (starting) chamber, to simulate the mechanical properties of the host sites. Every test has its own built-in control, so that results are always expressed as a preference index relative to this standard. There is a built-in method of gauging the overall responsiveness of the parasitoids and the possibility of the monitoring of their rate of entering the traps. These factors, combined with simplicity and economy of design and construction, ease of counting trapped parasitoids, unattended operation, absence of any need for a constant air flow and re-usability after cleaning, make the sandwich choicer "user-friendly."

While using the sandwich choicer, it was noticed that some individual *S. endius* (almost all female) were able to negotiate the adhesive coated trap dishes and enter the lure dishes. For some, an element of chance is probably involved. However, it is also possible that others of these pioneering individuals represent a sub-population especially adept at negotiating sticky substrates. If this is so, the sandwich choicer, or similar "obstacle race" device, might be used to select for individuals that could be used to breed biotypes that are better adapted to certain bio-control situations.

ACKNOWLEDGMENTS

I thank the University of Queensland for the provision of Special Studies Programme leave and Drs. Gary Mount, Dick Patterson and Phil Morgan of the United States Department of Agriculture, Agricultural Research Service, Insects Affecting Man and Animals Research Laboratory, Gainesville for helpful advice and provision of facilities and Mrs. Helen Amrhein and Mrs. Leslie Ellis for excellent secretarial support.

REFERENCES CITED

Dethier, V. G. 1947. Chemical Insect Attractants and Repellants. Blakiston, Philadelphia. p. 289.

Hogsette, J. A. and J. F. Butler. 1981. Field and laboratory devices for manipulation of pupal parasites, Status of Biological Control of Filth Flies. USDA-SEA, Gainesville, Florida. pp. 90-94.

Jacobi, E. F. 1939. Ueber Lebensweise, Auffinden des Wirtes und Regulierung des Individuenzahl von *Mormoniella vitripennis* Walker. Arch. neerl. Zool. 3:197-282.

Johnson, T. H. and O. W. Tiegs. 1922. What part can chalcid wasps play in controlling Australian sheep maggot flies? Queensland Agr. J. 17:128-131.

108

Knipling, E. F. and H. W. Brydon. 1966. Suppression of dung-inhabiting fly populations by pupal parasites. Ann. Entomol. Soc. Am. 59:638-651.

Legner, E.F. and E. J. Detrick. 1972. Innundation with parasitic insects to control filth breeding flies in California. Proc. 40th Ann. Conf. Calif. Mosquito Control Assoc.Inc. pp. 228-230.

Miller, D. 1927. Parasitic control of sheep maggot flies. New Zealand J. Agr. 34:1-4.

Morgan, P. B., R. S. Patterson, G. C. La Brecque, D. E. Weidhaas, and A. Benton. 1975a. Suppression of a field population of houseflies with *Spalangia endius*. Science 189:388-389.

Morgan, P. B., R. S. Patterson, G. C. La Brecque, D. E. Weidhaas, A. Benton, and T. Whitfield. 1975b. Rearing and release of the housefly pupal parasite *Spalangia endius* Walker. Env. Entomol. 4:609-611.

Morgan, P. B., R. S. Patterson, and G. C. La Brecque. 1976a. Host-parasitoid relationship of the housefly, *Musca domestica* L., and the protelean parasitoid, *Spalangia endius* Walker. J. Kansas Entomol. Soc. 49:483-488.

_____. 1976b. Controlling houseflies at a dairy installation by releasing a protelean parasitoid *Spalangia endius* (Hymenoptera: Pteromalidae). J. Georgia Entomol.Soc. 11:39- 43.

Morgan, P. B., G. C. La Brecque, and R. S. Patterson. 1978. Mass culturing the microhymenopteran parasite *Spalangia endius* (Hymenoptera: Pteromalidae). J. Med. Entomol. 14:671-673.

Morgan, P. B., D. E. Weidhaas, and R. S. Patterson. 1981. Programmed releases of *Spalangia endius* and *Muscidifurax raptor* (Hymenoptera: Pteromalidae) against estimated populations of *Musca domestica* (Diptera: Muscidae). J. Med. Entomol. 18:158-166.

Mourier, H. 1972. Release of native parasitoids of houseflies on Danish farms. Vidensk. Medd. Dan. Naturahist. Foren. Khobehavn 135:129-137.

Murphy, S. T. 1982. Host finding behavior of some Hymenopteran parasitoids of *Musca domestica*. Ann. Appl. Biol. 101:148-151.

Petersen, J. J., J. A. Meyer, D. A. Stage, and P. B. Morgan. 1983. Evaluation of sequential releases of *Spalangia endius* (Hymenoptera: Pteromalidae) for the control of houseflies and stable flies (Diptera: Muscidae) associated with livestock in Eastern Nebraska. J. Econ. Entom. 76:283-286.

Rice, M. J. 1989. The sensory physiology of pest fruit flies: conspectus and prospectus. In: Robinson, A.S. and Hooper, G. (eds.) Fruit Flies, Their Biology, Natural Enemies and Control. Elsevier, Amsterdam. pp. 249-272.

Stafford, K. C., C. W. Pitts, and T. L. Webb. 1984. Olfactometer studies of host seeking by the parasitoid *Spalangia endius* Walker. Env. Entomol. 13:228-231.

Varley, G. C. and R. L. Edwards. 1953. An olfactometer for observing the behavior of small animals. Nature, Lond. 171:789-790.

Vett, L. E. M., J. C. Van Lenteren, M. Heymens, and E. Meelis. 1983. An airflow olfactometer measuring olfactory responses of hymenopterous parasitoids and other small insects. Physiol. Entom. 8:97-106.

Vinson, S. B. 1976. Host selection by insect parasitoids. Ann. Rev. Entomol. 21:109-133.

Weidhaas, D. E., D. G. Haile, P. B. Morgan, and G. C. La Brecque. 1977. A model to simulate control of houseflies with a pupal parasite, *Spalangia endius*. Env. Entomol. 6:589-600.

Wylie, H. G. 1958. Factors that affect host finding by *Nasonia vitripennis* (Walk.) (Hymenoptera: Pteromalidae.) Can. Entomol. 90:597-608.

9. Biological Control of Muscoid Flies in Easter Island

Renato Ripa S.

In spite of the physical isolation of Easter Island many pests have arrived during the last 50 years, causing serious concern to the resident population. One of the problems originated from several species of flies which caused severe discomfort to residents, tourists, as well as to domestic animals. Considering the seriousness of the problem, the local authorities decided in 1981 to initiate a project which included the study and control of these flies. Initially, assistance was received from the Insects Affecting Man and Animals Research Laboratory, Gainesville, through Dr. R. S. Patterson who surveyed the problem in 1982 (Ripa, 1986). Results showed almost no natural enemies suggesting that a biological control program could be useful in reducing fly numbers. Besides, this would be highly suitable to a fragile island ecosystem.

PARASITE PRODUCTION, RELEASE AND EVALUATION METHOD

The project was started in 1982. Several species of beneficial insects were sent to Easter Island (Table 1). These were reared at the Subestación Experimental de Control Biológico La Cruz and shipped weekly to the Island. Most of the species were received initially from the aforementioned laboratory in Gainesville. The pupal parasites were placed in the field inside mesh bags. These were hung protected from direct sun near to larval development media e.g., cattle, caged-layer farm, refuse heap, etc. The action of the natural enemies on the flies was evaluated on seven occasions, sampling different fly development media each time. The effect of the pupal parasites was evaluated by retrieving the pupae from soil

111

Table 1. Natural enemies of muscoid flies released in Easter Island, Chile.

Species	Year	Total No. Released
Spalangia endius 4 strains (Gainesville Fl. High Spring Fl., Thailand, New Zealand)	1982/88	11,793,649
Spalangia cameroni	1983/88	3,539,720
Muscidifurax raptor 2 strains (Chile, Trendelburg Germany	1983/88	4,625,360
M. zaraptor	1986/88	315,000
Pachycrepoideus vindemmiae	1985/88	773,000
Creophilus erythrocephalus	1984/85	850
C. maxillosus	1985	50

near or beneath cattle dung, with a series of sieves. Similarly pupae were recovered from the soil of the municipal dump, poultry farm and Anakena beach garbage. The pupae were placed in vials and stored at 25°C until emergence of adult flies or parasites. After twenty-five days the unemerged pupae were dissected in search of the mortality factor. In addition, the number of predators was also recorded per area or dung pad.

RESULTS

Several species of flies were identified from the samples; these are indicated in Table 2. Most of the problem arises from cow

Table 2. Fly species observed on Easter Island.

Muscidae

1. *Musca domestica* L.
2. *Fannia* sp.
3. *Ophyra aenescens*
4. *Stomoxys calcitrans* (L.)

Calliphoridae

5. *Phaenicia sericata* (Meigen)
6. *Paralucilia fulvicrura*
7. *Sarconecia chlorogaster* (Wiedmann)

Sarcophagidae

8. *Sarcophaga haemorrhoidalis* (Fallen)
9. *Sarcophaga* sp.

dung where flies oviposit and the larvae develop. No studies have been carried out concerning the population fluctuation of larvae and adults. However, larval population in the dung varies during the year and between sites on the Island. Probably this depends on pasture quality, rainfall, wind and other weather factors which affect dehydration and suitability of the media.

The first natural enemy survey before the introduction program showed no pupal parasites (Patterson and Ripa, 1982). It was observed that the only predator was a cricket, *Teleogryllus oceanicus* (Le Guill). This cricket was present in relatively high numbers, sheltered under the numerous volcanic stones which cover the Island. Preliminary tests showed that the adult cricket may consume up to ten fly larvae in a twenty-four hour period. Interspecific competence occurred with the only species of dung beetle, *Aphodiur lividus* Oliv., present on the Island. The population of this beetle is highly variable. Some dung pads were colonized by more than one hundred larvae, others showed a

Table 3. Parasitism of field collected muscoid pupae. Easter Island, Chile.

Pupal media	March 1985	Nov. 1985	May 1986	March 1987	Oct. 1987	June 1988
Cattle droppings						
No. pupae collected	*147(2)	258(3)	12(1)	307(2)	741(3)	75(2)
% adult emergence	75.51	39.53	0	59.28	40.48	58.66
% parasitism	15.64	31.0	8.33	3.90	27.39	10.66
Garbage						
No. pupae collected	1,214(3)	108(2)	157(1)	252(1)	--	53(1)
% adult emergence	59,88	25.0	74.52	51.98	--	56.6
% parasitism	33,77	28.70	0	18.25	--	30.18
Poultry droppings						
No. pupae collected	2.871(1)	174(2)	223(1)	197 (1)	346 (1)	--
% adult emergence	60.26	28.73	57.40	78.68	94.21	--
% parasitism	0.90	28.26	8.97	0.50	3.7	--

* Value between parentheses represents number of sites where sampling was carried out.

negligible number on were totally absent. Dung pads highly colonized by the beetle tended to show a reduced population or even an absence of fly larvae. However, the overall impact of the beetle on the fly problem appeared to be of limited value.

The second survey in March 1983, six months after initiation of periodic releases showed that pupae collected in the poultry farm, Municipal dump and Vaitea cattle pasture ground were not parasitized. Just two pupae collected under a dust bin showed the emergence hole of a parasite. This extremely low level of parasitism might reflect the short period between the initiation of periodic releases and the evaluation. The samplings from 1985 onwards are indicated in Table 3. As each site was sampled during approximately forty-five minutes the number of pupae collected reflects as well a crude estimate of their abundance. On several occasions, the annoying population of adults experienced during sampling did not reflect the low density of pupae found in the soil, suggesting that the adults were pooled from the surroundings or originated from other sources in the area. Parasitism in general fluctuated considerably between sampling sites and the years. Moreover, in some of the samples the limited number of pupae collected probably did not reflect an accurate level of parasitism. It is interesting to note that occasionally the poultry farm manure and soil was heavily treated with pesticides including chlorinated insecticides. This may account for the low parasitism of *Musca domestica* pupae on most of the surveys. It was observed during the March 1987 sampling that the population of adult flies was exceptionally high. A long drought period had occurred from November 1986 to mid February 1987. The first rain after the drought period produced a flush of young growth of gramminae. The feces originating from feeding on tender growth seemed to be very suitable for fly development. Very high numbers of adults were observed after about twenty-five or more days posterior to the flush of tender growth. The low population of flies during a drought period limited the abundance of parasites which were not capable to increase their population with the same rate as the flies. This probably explains the low level of parasitism observed from cattle dung pupae in the March 1987 sampling (Table 3).

The comparison between released species, Table 1, versus the species recovered in the sample, Table 4, shows that *Spalangia endius* is the most frequent and successful species of the released ones. *Muscidifurax raptor* was recovered on only one occasion, March 1985, which coincides with the end of the releasing period of the Chilean strain, probably suggesting that the Trendelburg strain is

Table 4. Hymenopterous parasites reared from muscoid pupae, Easter Island, Chile.

Species	Number	Origin
Tachinaephagus zealandicus	467	Unknown probably arrived 1984
Spalangia endius	357	Introduced
Aphaere laeriuscasta	44	Unknown probably arrived 1986/87
Spalangia sp.	22	Probably either *endius* or *cameroni*
Muscidifurax raptor	12	Introduced

even less active in the Easter Island environment. *M. zaraptor* and *Pachycrepoideus vindemmiae* was not recovered in the samples suggesting that these species are not suitable. It is also surprising that a different species, *Tachinaephagus zealandicus*, identified by Dr. Luis De Santis was recovered mainly from pupae of the Anakena beach garbage. This species was not deliberately introduced and probably arrived before 1985. It was reared from sarcophagid and calliphorid pupae.

With respect to predators, the staphilinds *Creophilus erytrocephalus* and *C. maxillosus* were collected in poultry farms near to Quillota, Chile. The adults were rid from phoretic mites by submerging them for a few seconds in cyhexatin 0.2 g a.i./lt., and then repeated again after 48 h. Just 900 specimens were sent between 1984 and 1985. Establishment and active breeding of *C. erytrocephalus* was observed in the November 1985 sampling; before that, its presence was recorded during September of that year by G. Velasco (personal communication). This is less than one year after the initial releases. This staphilind seems to cause an important mortality of fly larvae in caged-hen manure, since an increasing number of adults were recorded during the following years after establishment (Table 5). A maximum of forty adults per square meter of dung were recorded. The adults were also observed preying on the neonatal adult flies as they emerged. It is interesting to note that *C. erytrocephalus* is a good flier. This helped the staphilinid to colonize most of the Island during the first year, including places more than fifteen km from the release point. The impact of this staphilind in other fly breeding media seems to be less important. The second staphilind, *C. maxillosus* was not recovered. Correspondingly, it is also less frequent in poultry manure in Central Chile.

During the four years of the study sampling sites had to be changed according to the availability of fly rearing media. Cattle were shifted to different fields throughout the Island. This presumably changed the environment and possibly the influence on the parasites and their control. The poultry farm was closed in 1987 and with it probably a continuous source of the staphilinid. Unfortunately there is no adult fly population record before and during the study. There is however a general appreciation in the community that the fly problem on the Island has decreased considerably. Parasites control an important part of the population, through parasitization as observed in Table 3. The lower than expected emergence of adult flies suggests that dudding by the adult parasite is also an important mortality factor which should be considered in the global effect of pupal parasites (Petersen, 1986). The removal of the pupae while they are still acceptable for further attack (Simmonds, 1949), generally underestimates the real impact of parasites. Considering these effects probably explains the lower population of adult flies observed.

SUMMARY

A biological control program of filth flies was initiated in Easter Island from 1982 onwards. Several natural enemies were introduced from mainland Chile. The following species of pupal parasites were sent between 1982 and 1988, *Spalangia endius, S. cameroni, Muscidifurax raptor, M. zaraptor* and *Pachycrepoideus vindemmiae*. Establishment was observed with *S. endius* and in one occasion with *M. raptor*. Seven evaluations between 1983 and 1988 indicated a variable parasitism, with maximum near to thirty per cent. However, dudding and host removal while still acceptable for further attack may account for even higher mortality. Two species of staphilind predators were sent in 1984 and 1985. *Creophilus*

Table 5. Abundance of *Creophilus/erytrocephalus*, Easter Island, Chile.

Type of habitat	March 1985	Nov. 1985	May 1986	March 1987	Oct. 1987	June 1988
Cattle droppings						
No. of dung pads searched	0	150	25	16	45	22
No. of insects observed		13 larvae	not re-covered	not re-covered	11 larvae 7 adults	1 larvae
Garbage	0	present	present	not re-covered	--	not re-covered
Poultry Farm						
No. of adults per m² of droppings	0	6	17	20	4	--

erytrocephalus F. became establishment reaching high population levels in poultry manure, affording good control of *Musca domestica.* However, the second species *C. maxillosus* L. was not recovered. There is a general assent that the muscoid fly population is now considerably lower and inhabitants and tourists are less annoyed.

REFERENCES CITED

Patterson, R. S. and Ripa, R. 1982. El problema de las moscas en Isla de Pascua (Chile) y posibles formas de suprimirlas. Report. Subestacion Experimental La Cruz (INIA), Chile. 26 pp.

Petersen, J. J. 1986. Evaluating the impact of pteromalid parasites on filth fly populations associated with confined livestock installations, Biological Control of Muscoid Flies. R. S. Patterson and D. A. Rutz, eds. Miscellaneous Publication. Ent.Soc. of Amer. pp. 52-56.

Ripa, R. 1986. Survey and use of biological control agents on Easter Island and in Chile. Biological Control of Muscoid Flies. R. S. Patterson and D. A. Rutz, eds. Miscellaneous Publication. Ent.Soc. of Amer. pp. 39-44.

Simmonds, F. J. 1949. Some difficulties in determining by means of field samples the true value of parasitic control. Bull. Entomol. Res. 39: 435-440.

10. Filth Flies and Their Potential Natural Enemies in Poultry Production Systems in the Philippines

L. M. Rueda, C. T. Hugo and M. B. Zipagan

ABSTRACT

An extensive survey of house fly and other filth flies, and their arthropod predators and parasites associated with broiler and caged-layer poultry production facilities was conducted in selected areas of Luzon, Philippines. About eleven species of common filth flies (Diptera) belonging to eight genera and four families (Muscidae, Calliphoridae, Sarcophagidae, and Stratiomyidae) were collected and identified. Eight species of parasitic Hymenoptera, belonging to six genera and three families (Pteromalidae, Chalcididae and Diapriidae) were found attacking pupae of *Musca domestica* L. and other filth flies. About thirty-five species of arthropod predators and scavengers, classified under fourteen families and five orders were recovered in poultry manure. The relative abundance of the common parasites and predators was also determined. Rates of predation on house fly eggs and first instar larvae were determined for selected insect predators.

INTRODUCTION

Poultry production is an important component of Philippine agriculture. As a result of increased financial support from government and private entities coupled with high demand for poultry meat, eggs and related by-products, many farmers are encouraged to shift from backyard to commercial poultry production. This results in more manure production and an increase in filth fly problems associated with manure accumulation. Filth flies, particularly the house fly (*Musca domestica* L.) and other muscoid flies, are a serious threat to poultry production. They cause annoyance to poultry, human caretakers, as well as other people

121

residing around or near the poultry farms. They are also potential carriers of various microbial and related diseases of poultry, livestock and human beings. With the high costs of synthetic chemical pesticides, in addition to problems associated with their injudicious usage, it is necessary to find an alternative strategy to manage fly pest population. Parasites and predators could be integrated with cultural (e.g., manure management, etc.) and chemical methods to develop a program for managing the fly problems in poultry production facilities (Legner et al. 1975, Axtell 1981, 1986a,b, Axtell and Rutz 1986). In the Philippines, insufficient information about the identity and potential of these biological control agents hinders their full utilization as an important component of a fly management program.

Published reports on filth flies and associated arthropods in poultry manure in the Philippines are few and scattered. Rueda (1985a,b) indicated taxonomically about 30 species of blowflies (Calliphoridae), some of which are commonly found in poultry farms. Cabrera & Rozeboom (1956) provided adult descriptions and notes on the distribution and habits of several species of *Musca*. Baltazar (1966) listed two species of pteromalid wasps parasitizing fly pupae. In 1986, Rueda reported six additional hymenopterous parasites of muscoid fly pupae associated with poultry manure. Arellano and Rueda (1988) investigated various aspects of the biology of *Spalangia endius* (Walker). A brief list of mites associated with poultry manure was provided by Corpuz-Raros et al. (1988).

This study was initiated to determine the kind of common filth flies and their potential natural enemies in poultry production facilities. Relative abundance, and rates of parasitism and predation of selected natural enemies were determined.

MATERIALS AND METHODS

Sample Sites. The survey was conducted on about fourteen poultry farms located in two provinces (i.e. Laguna and Batangas) of Luzon, Philippines. Broiler poultry houses were open sided structures, 90-110 m long by 8-10 m wide with capacities of 5,000-7,000 birds. Slated platforms (two m above the dirt floor) cover the entire area of the house along each wall. Feeders and waterers were placed over the slats, and manure accumulated underneath. Caged-layer houses were open sided structures (40-80 m long by four m wide with 2,000-3,000 bird capacity) with one row of two-tiered stairstep cages, three to four birds per cage, suspended above a dirt floor, and running the length of the house along each side of concrete walkways.

Sampling and Identification. Adult filth flies in and around poultry houses were collected using sweep nets and/or baited jug traps. These traps (Burg & Axtell 1984) were 3.8 liter plastic jugs with four holes (five cm diameter) cut in the upper part of the sides to allow entrance of the flies which were attracted to thirty grams of fly bait (ten parts sugar, two parts rice bran, one part azamethiphos 50%) placed on the bottom of the jug. The traps (two per house) were suspended with wire (thirty-five cm length) from the roof support on both ends of the house. The traps were left for two days and the species of flies collected were identified. Sweepings (ten sweeps/house) were made along both sides of the house. Flies attracted to dead chickens were also collected using a sweep net and identified.

Predaceous and scavenging arthropods were collected from manure samples. Each sample, about four liters of manure, was gathered in a single location with a hand towel and placed in a six-liter plastic container lined with a plastic bag. The top edge of the bag was folded down over the rim of the container before the cover was put in to allow air exchange and to prevent excessive condensation. The manure samples were placed in Tulgreen funnels upon a wirescreen insert (thirty cm diameter, 6.3 mesh openings). A 50-w bulb on top of each funnel served as the heat light source to drive the arthropods downward into a 190 ml jar containing 70% ethyl alcohol. Samples were left to dry on the funnels for seventy-two hours. After the extraction period, the contents of the jar were transferred to screw-capped jars where they were held for counting using a binocular microscope. Since Tulgreen funnels were effective only in sampling motile organisms, dried manure was hand sorted under a glass magnifier to recover dormant stages of arthropods.

For parasitic Hymenoptera, two techniques were used to survey and monitor their populations in each of the poultry farms. They were: 1) collection of naturally occurring fly pupae in the poultry manure and rearing them in the laboratory for parasite development and emergence, and 2) pupal bag technique, which involved using fourteen-mesh screen bags each containing thirty laboratory-reared, less than twenty-four-hour old house fly pupae. The bag was inserted in the manure at three to five cm depth and exposed for one week. The pupae were retrieved and placed in plastic vials in the laboratory for parasite development and emergence. In each poultry farm, about thirty pupal bags were positioned in the manure, and four two-liter manure samples (for naturally occurring fly pupae) were collected.

The data collected weekly were combined for six months with four sampling periods per month. The relative abundance of each species of predatory and scavenging arthropods was calculated based on the total number of adults and immatures collected per manure sample over the entire six-month period. The following index for the number of individual adults and immatures (i.e. larvae for insects, protonymphs and deutonymphs for mites) per manure sample was used: + = 1-10, ++ = 11-50, +++ = 51-100, ++++ = 101 or more. The percentage of parasitism by hymenopterous species was calculated for the six-month period as the percentage of the exposed (sentinel) fly pupae or naturally occurring pupae which were recovered intact and from which adult parasites emerged (assuming one parasite developed per pupa).

Predation Rates. Beetle, anthocorid, earwig and mite predators were collected in 1985 from poultry houses in Laguna province, Luzon and maintained in screened-topped 3.8-liter plastic containers (sixteen cm height, twenty-one cm diameter) filled up to eight cm depth with 1:0.5 mixture of fresh poultry manure and rice bran. Frozen house fly eggs were added to the cultures every two to three days, and predators were transferred to fresh medium every two weeks.

Insect predators were removed from the cultures and held without prey for twenty-four hours prior to predation tests. Each female adult insect (about two to three days old) was placed in a closed six-dram plastic "snap cap" vial and held at twenty-seven to twenty-nine degrees C. Each vial was lined with moistened blotting paper on which there were 300 house fly eggs (one to three hours old). After successive twenty-four-hour intervals, the insect was transferred to another vial containing fresh eggs and moistened blotting paper. Each insect was tested for five successive twenty-four-hour intervals. Egg and larval destruction was determined daily by counting the number of intact first instar larvae and the number of non-punctured intact egg chorions per vial. The difference between these counts and initial number of eggs exposed (300) was the number of eggs and first instar larvae destroyed by the predator. Vials lined with moistened blotting paper with house fly eggs without predators were used as controls. About twenty adults per predator species were evaluated for predation. Two separate tests for each species were conducted.

The same procedure for evaluating predation of insects was followed for mites with minor modifications. Protonymphs and deutonymphs in addition to adult female mites were included in the predation tests. About twenty house fly eggs (one to three hours old) were placed in moistened blotting paper in a three-dram plastic "snap cap" vial. One mite was transferred to another vial containing fresh eggs and moistened blotting paper. Egg and first instar larval destruction was determined at twelve-hour intervals. Protonymps and deutonymphs, and adult mites were tested for predation for two and six successive twelve-hour intervals, respectively. Each test was conducted twice, using fifteen predators (one predator per vial with fly eggs) in each developmental stage of mite species, plus controls.

Analysis of variance (ANOVA) was used to identify significant predation rates of predators. Differences in treatment means were judged at P = 0.05 using Duncan's multiple range test (Duncan 1955).

RESULTS AND DISCUSSION

<u>Species Composition and Abundance</u>. About eleven species of common filth flies, under four families of Diptera were collected in poultry farms. These include *Musca domestica*, *M. sorbens* Weidemann, *Orthellia indica* Robineau-Desvoidy, *Fannia* sp. (Family Muscidae); *Chrysomya megacephala* (Fabricius), *C. rufifacies* (Macquart), *Lucilia porphyrina* (Walker), *Lucilia papuensis* Macquart, *Phaenicia cuprina* (Weidemann) (Family Calliphoridae); *Parasarcophaga albiceps* (Meigen) (Family Sarcophagidae); and *Hermetia illuscens* (L.) (Family Stratiomyidae). *M. domestica* was the most common fly species, followed by *C. megacephala* and *H. illuscens* in all farms surveyed. *C. megacephala*, together with other calliphorid and sarcophagid species, were abundant particularly in farms with dead chickens and small animals in and around the poultry houses. During early morning and late afternoon, *C. megacephala* and other blowflies aggregated mostly on vegetation around or near the poultry houses. Although house flies were also found resting on vegetation around or near the houses, most of them rested on the walls, ceilings, gutters, beams and even on feed and water troughs in the poultry house during heavy infestations. The larval activity of *C. megacephala*, like *H. illuscens*, resulted in the semi-liquid condition of the manure. This provides extreme difficulties in removing

manure from the poultry house and possibly reduce the plant fertilizer qualities of the manure (Axtell & Edwards 1970).

The species composition and relative abundance of arthropod predators and scavengers, collected from poultry manure are presented in Table 1. *Phacophalus tricolor* Krantz (Staphylinidae), *Carcinops* sp. (Histeridae) and *Alphitobius diaperinus* (Panzer) (Tenebrionidae) were consistently abundant species in all farms surveyed. Other beetles, such as *Eleusis kraatzi* Fauvel (Staphylinidae), *Dactylosternum abdominale* (Fabricius), *Spharaedium* sp. (Hydrophilidae), *Onthophagus* sp. and *Trox* sp. (Scarabaeidae), were also common in most poultry farms. *Euborellia philippinensis* Srivastava (Anisolabidae, Dermaptera) was fairly common in poultry manure. Among the mites, species under families Macrochelidae (i.e. *Macrocheles muscaedomesticae* (Scopoli), *M. merdarius* Berlese, *Glyptholaspis confusa* (Foa), *Glyptholaspis* sp.), Parasitidae (one unidentified species) and Uropodidae (one unidentified species) were abundant in poultry manure.

Among the parasitic Hymenoptera, about eight species under three families were recovered from naturally occurring pupae and sentinel pupae of the house fly in poultry manure (Table 2). These include *Pachycrepoideus vindemmiae* (Rondani), *Spalangia endius* (Walker), *S. cameroni* Perkins, *S. nigroaenea* Curtis, *Muscidifurax raptor* (Girault and Sanders), *Nasonia vitripennis* (Rondani) (Pteromalidae); *Dirhinus himalayanus* (Westwood) (Chalcididae); and *Trichopria* sp. (Diapriidae). *P. vindemmiae* was the dominant species parasitizing house fly pupae (both naturally occurring and sentinel) in poultry farms. It was followed by *S. endius, S. cameroni, Trichopria* sp. and *D. himalayanus. M. raptor*, an abundant species in North Carolina (Rueda & Axtell 1985a,b,c, 1987, Rutz & Axtell 1980), was not commonly recovered from poultry farms in Luzon provinces (i.e. Batangas and Laguna). All parasite species, except *N. vitripennis* and *M. raptor*, occurred throughout the sampling months (May to October) with peak of abundance in July. Parasitism rates of house fly pupae averaged about 15.7 and 50.2%, respectively, using pupal bag technique and naturally collected fly pupal samples.

From naturally occurring pupae of *C. megacephala* in poultry manure, about four species (i.e. *S. endius, S. cameroni, P. vindemmiae* and *D. himalayanus)* were recovered. *S. endius* (80.3%) was the dominant species parasitizing *C. megacephala* pupae, followed by *P. vindemmiae* (11.0%), *S. cameroni* (6.0%) and *D. himalayanus* (1.7%). Parasitism rate of naturally occurring

Table 1. Predatory and scavenging arthropods found in manure on poultry farms in Luzon, Philippines.

Species	Relative abundance[1]
Insecta	
Coleoptera	
Staphylinidae	
Aleochara puberula Klug	++
Eleusis kraatzi Fauvel	+++
Phacophalus tricolor Krantz	++++
Staphylinids (3 spp.)	+++
Histeridae	
Carcinops sp.	++++
Histerids (3 spp.)	++
Tenebrionidae	
Paloruz ratzeburgii (Wissemann)	+
Mesomorphus villiger Blanchard	+
Alphitobius diaperinus Panzer	++++
Tenebrionids (1 sp.)	+
Hydrophilidae	
Dactylosternum abdominale (Fabricius)	++
Spharaedium sp.	++
Hydrophilids (1 sp.)	++
Scarabaeidae	
Onthophagus sp.	++
Trox sp.	++
Scarabaeids (1 sp.)	+
Dermestidae	
Dermestes sp.	+
Carabidae (1 sp.)	+++
Dermaptera	
Anisolabidae	
Euborellia philippinensis (Srivastava)	+++
Euborellia sp.	++
Hemiptera	
Anthocoridae	
Xylocoris sp.	++
Reduviidae (1 sp.)	+

(Continued)

Table 1.1 (Cont.) Predatory and scavenging arthropods found in manure on poultry farms in Luzon, Philippines.

Species	Relative abundance[1]
Hymenoptera	
Formicidae (3 spp.)	++
Arachnida	
Parasitiformes	
Macrochelidae	
Macrocheles muscaedomesticae (Scopoli)	++++
M. merdarius Berlese	++++
Glyptholaspis confusa (Foa)	++++
Glyptholaspis sp.	+++
Parasitidae (1 sp.)	++++
Uropodidae (1 sp.)	++++

[1] Based on the mean number of adults and immatures collected per four to five liter manure sample (438 samples from fourteen farms) during May to October, 1986.

pupae of *C. megacephala* was 46.2% during the sampling months (May to October).

Previous surveys (Rueda and Zipagan, unpublished data) in various parts of the Philippines (i.e. North and Central Luzon, Bicol Region, Cebu, Leyte, Panay, Palawan, Mindoro, Davao, Cagayan de Oro and other Mindanao provinces) indicated the presence of these eight parasite species in most areas, with *P. vindemmiae* and *S. endius* having the highest rate of parasitism on naturally occurring house fly pupae in poultry manure.

Predation Rates. The rates of house fly egg and first instar destruction by female adult beetles, anthocorid bugs and anisolabid earwigs are shown in Table 3. Adult females of *A. puberula* destroyed the highest numbers of egg and first instar larvae (about 255 per beetle adult per day). *E. philippinensis*, the second voracious predator in the tests, destroyed about 177 eggs and first instar larvae per earwig per day. *P. tricolor*, *D. abdominale*, *E. kraatzi*, *Spharaedium* sp. and *Onthophagus* sp. destroyed about 88, 79.6, 69.5, 91 and 71 eggs and first instar larvae per beetle per day, respectively. *Carcinops* sp. and *Xylocoris* sp., the less voracious

Table 2. Relative abundance of parasitic Hymenoptera that emerged from house fly pupae exposed in mesh bags or naturally occurring pupae in poultry manure in houses on five farms in Laguna, Luzon, Philippines.

Parasite species	Relative abundance(%)[1]	
	Sentinel pupae	Naturally occurring pupae
Pachycrepodeus vindemmiae	64.3	44.1
Spalangia endius	20.4	29.8
S. cameroni	8.3	3.3
S. nigroaenea	0.2	0.7
Muscidifurax raptor	0.6	1.1
Nasonia vitripennis	0.6	4.2
Dirhinus himalayanus	0.8	7.6
Trichopria sp.	5.9	9.2
Total parasites recovered (n)	1499.0	551.0
Total pupae collected intact (n)[2]	9458.0	1097.0
Parasitized pupae (%)[3]	15.7	50.2

[1] Relative abundance means are based on the total number of parasites collected during the six-month period from sentinel or naturally occurring pupae which were recovered intact in each of the different poultry farms (thirty pupae per bag, fifteen bags per farm per week).

[2] Total number of sentinel or naturally occurring house fly pupae recovered intact.

[3] Percentage of sentinel or naturally occurring house fly pupae recovered intact from which adult parasites emerged. Parasitism means were calculated from the total number of parasites recovered and the total number of sentinel or naturally occurring pupae recovered intact during the six-month period (May-October, 1986).

Table 3. Rates of predation of house fly eggs and first instar larvae by adult beetles, anthocorid bugs and anisolabid earwigs.

Species	Predation Rate[1] Mean (S.E.)	Range
Coleoptera		
Staphylinidae		
Phacophalus tricolor	88.0 (9.0) c	69-101
Eleusis kraatzi	69.5 (17.3)c	24-100
Aleochara puberula	255.0 (16.1)a	161-298
Dactylosternum abdominale	79.6 (16.1)c	39-100
Spharaedium sp.	91.0 (9.4) c	90-102
Histeridae		
Carcinops sp.	37.2 (4.2) d	16-50
Scarabaeidae		
Onthophagus sp.	79.0 (6.0) c	21-94
Hemiptera		
Anthocoridae		
Xylocoris sp.	38.1 (4.6) d	11-76
Dermaptera		
Anisolabidae		
Euborellia philippinensis	117.2 (12.3)b	95-183

[1] Number of house fly eggs and first instar larvae destroyed per adult female predator in 5 days; corrected for control mortality by Abbott's (1925) formula. Means followed by the same letter are not significantly different at the 5% level (Duncan's (1955) multiple range test).

predators, destroyed about 37.2 and 38.1 eggs per adult beetle per day, respectively.

Numerous beetles, particularly staphylinids and histerids, were considered important predators on fly eggs and larvae in poultry manure (Legner 1971, Peck 1969, Peck & Anderson 1969, Pfeiffer & Axtell 1980). *Carcinops pumilio* (Erichson) adults consumed about thirteen fly eggs per day (Morgan et al. 1983) to more than a hundred (Geden et al. 1988) depending upon the prey density, and previous feeding history of the predators. Insufficient published reports on the predation rates of the insect species

Table 4. Rates of predation on house fly eggs and first instar larvae by macrochelid mites.

Species	Predation Rate[1] Mean (S.E.)	Range
Macrocheles muscaedomesticae		
Protonymph	0.7 (0.4)a	0-1
Deutonymph	6.0 (0.6)c	5-11
Adult female	9.7 (0.9)c	6-13
M. merdarius		
Protonymph	0.6 (0.3)a	0-1
Deutonymph	1.6 (0.4)b	1-3
Adult female	3.7 (0.6)b	2-6
Glyptholaspis confusa		
Protonymph	0.8 (0.3)a	0-1
Deutonymph	5.8 (0.8)c	3-9
Adult female	8.7 (0.9)c	4-13

[1] Number of house fly eggs and first instar larvae destroyed per individual predator in 3 days for adult female mite, 1 day for protonymph and deutonymph; corrected for control mortality by Abbott's (1925) formula. Means followed by the same letter are not significantly different at the 5% level (Duncan's (1955) multiple range test).

included in this study make it impossible to compare species/strain differences.

In three mite species tested, the protonymphs destroyed fewer numbers of house fly immatures than deutonymphs and adults (Table 4). There was no significant difference between the numbers of house fly eggs and first instar larvae destroyed by deutonymphs and adults in each species. Deutonymphs and adults of *M. muscaedomesticae* and *G. confusa* destroyed a higher number of prey than those of *M. merdarius*. Predation rates of *M. muscaedomesticae* (reviewed by Axtell 1969) vary greatly depending upon the type of subtrates used (i.e. animal manure and fly rearing media vs. absorbent cotton and paper) (Rodriguez & Wade 1961, Axtell 1961, O'Donnel & Axtell 1965, Willis & Axtell 1968, Peck 1969, Toyoma & Ikeda 1976), predator density (Geden et al. 1988), presence of alternate prey (i.e. nematodes) (Ito 1973, 1977), and possibly type of strains. *G. confusa* adult females

132

destroyed about four to thirteen (average 8.7) house fly eggs and first instar larvae per mite per day, and these rates were almost similar to those obtained by Axtell (1961), Peck (1969) and Geden et al. (1988). *M. merdarius* adult females were relatively less efficient predators, as compared with *M. muscaedomesticae* and *G. confusa*.

Determination of species composition and abundance is one of the essential steps towards unlocking the complexities of the arthropod community structure in poultry manure. Preliminary laboratory studies on parasitism and predation rates of various species could give us some perceptions of their potential as biological control agents. These, however, must be supplemented by field studies to have more conclusive results. Numerous factors, i.e. temperature, prey density, manure habitat, alternative hosts, competition, etc. would definitely affect the natural rates of parasitism and predation of various species under field conditions. The results from our study provided some basic information of the species composition and abundance of common parasitic and predatory arthropods in poultry manure in the Philippines. Further studies to determine the effectiveness of major species of natural enemies as they are affected by various factors (biotic and abiotic) are necessary before they could profoundly be integrated in the poultry fly pest management programs.

ACKNOWLEDGMENTS

The assistance of G. Arellano and L. de Jesus is acknowledged. Thanks to the following for confirming identifications of the specimens: Dr. Z. Boucek, Chalcididae; Drs. G. W. Krantz & B. Halliday, Macrochelidae; Dr. V.P. Gapud, Anthocoridae; taxonomists in the British Museum (Natural History, Entomology), Coleoptera. We also thank Dr. R.C. Axtell for reviewing the manuscript. This research was supported in part by the University of the Philippines Basic Research Project No. 85-2, National Science and Technology Authority Project No. 8505 and International Foundation for Science (Sweden).

REFERENCES CITED

Abbott, W. S. 1925. A method for computing the effectiveness of an insecticide. J. Econ. Entomol. 18: 265-267.

Arellano, G. M. and L. M. Rueda. 1988. Biological study of the house fly pupal parasitoid, *Spalangia endius* (Walker) (Hymenoptera: Pteromalidae). Philipp. Entomologist 7(4):329-350.

Axtell, R. C. 1961. New records of North American Macrochelidae (Acarina: Mesostigmata) and their predation rates on the house fly. Ann. Entomol. Soc. Am. 54:748.

_____. 1969. Macrochelidae as biological control agents for synanthropic flies, pp. 401-416. In G.O. Evans (ed.), Proc. 2nd International Congress of Acarology. Akademiae Kiado, Budapest.

_____. 1981. Use of predators and parasites in filth fly IPM programs in poultry housing. pp. 26-43. In R.S. Patterson, P.G. Koehler, P.B. Morgan and R.L. Harris (eds.), Status of biological control of filth flies. U.S. Dept. of Agric. Res. Serv. SEA A 106.2:F64.

_____. 1986a. Fly management in poultry production: Cultural, biological, and chemical. Poultry Science 65:657-667.

_____. 1986b. Status and potential of biological control agents in livestock and poultry production systems, pp. 1-9. In R.S. Patterson and D.A. Rutz (eds.), Biological control of muscoid flies. Entomol. Soc. Am. Misc. Publ. No. 62.

Axtell, R. C. & T. D. Edwards. 1970. *Hermetia illuscens* control in poultry manure by larviciding. J. Econ. Entomol. 63:1786-1787.

Axtell, R. C. & D. A. Rutz. 1986. Role of parasites and predators as biological fly control agents in poultry production facilities, pp. 88-100. In R. S. Patterson and D. A. Rutz (eds.), Biological control of muscoid flies. Entomol. Soc. Am. Misc. Publ. No. 62.

Baltazar, C. R. 1966. A catalogue of Philippine Hymenoptera with bibliography 1958-1963. Pacific Insects Monograph 8. 453 pp.

Burg, J. G. & R. C. Axtell. 1984. Monitoring house fly, *Musca domestica* (Diptera: Muscidae), population in caged-layer poultry houses using a baited jug trap. Environ. Entomol. 13:1083-1090.

Cabrera, B. D. & L. E. Rozeboom. 1956. The flies of the genus *Musca* in the Philippines. Philipp. J. Sci. 85(4): 425 449.

Corpuz-Raros, L. H., G. C. Sabio & M. Velasco-Soriano. 1988. Mites associated with stored products, poultry houses and house dust in the Philippines. Philipp. Entomologist 7(3): 311-321.

134

Duncan, D. B. 1955. Multiple range and multiple F tests. Biometrics 11:1-41.

Geden, G. J., R. E. Stinner & R. C. Axtell. 1988. Predation by predators of the house fly in poultry manure: Effects of predator density, feeding history and interspecific interference and field conditions. Environ. Entomol. 17(2): 320-329.

Ito, Y. 1973. The effects of nematode feeding on the predatory efficiency for house fly eggs and reproduction rate of *Macrocheles muscaedomesticae* (Acarina: Mesostigmata). Jpn. Sanit. Zool. 23:209-213.

Legner, E. F. 1971. Some effects of the ambient arthropod complex on the density and potential parasitization of muscoid Diptera in poultry wastes. J. Econ. Entomol. 64:111-113.

Legner, E. F., W. F. Bowen, W. F. Rooney, W. D. Mckeen & G.W. Johnston. 1975. Integrated fly control on poultry ranches. Calif. Agric. 29(5):8-10.

Morgan, P. B., R. S. Patterson & D. E. Weidhaas. 1983. A life history study of *Carcinops pumilio* Erichson (Coleoptera: Histeridae). J. Ga. Entomol. Soc. 18:353-359.

O'Donnell, A. E. & R. C. Axtell. 1965. Predation by *Fuscuropoda vegetans* (Acarina: Uropodidae) on the house fly (*Musca domestica*). Ann. Entomol. Soc. Am. 58:403-404.

Peck, J. H. & J. R. Anderson. 1969. Arthropod predators of immature Diptera developing in poultry droppings in northern California. Part II. Laboratory studies on feeding behavior and predation potential of selected species. J. Med. Entomol. 6: 163-167.

Pfeiffer, D. G. & R. C. Axtell. 1980. Coleoptera of poultry manure in caged layer houses in North Carolina. Environ. Entomol. 9:21-28.

Rodriguez, J. G. & C. F. Wade. 1961. The nutrition of *Macrocheles muscaedomesticae* (Acarina: Macrochelidae) in relation to its predatory action on the house fly egg. Ann. Entomol. Soc. Am. 54:782-788.

Rueda, L. M. 1985a. Some Philippine blowflies (Diptera: Calliphoridae). I. Subfamily Calliphorinae. Philipp. Entomologist 6(3):307-358.

_____. 1985b. Some Philippine blowflies (Diptera: Calliphoridae). II. Subfamilies Chrysomyinae, Rhiniinae and Ameniinae. Philpp. Entomologist 6(4):362-390.

Rueda, L. M. 1986. Hymenopterous parasites of house fly (*Musca domestica* Linn.) and other filth flies in poultry production systems in the Philippines. International Foundation for Science (Sweden) Prov. Report No. 22:109-119.

Rueda, L. M. & R. C. Axtell. 1985a. Comparison of hymenopterous parasites of house fly, *Musca domestica* (Diptera: Muscidae), pupae in different livestock and poultry production systems. Environ. Entomol. 14:217-222.

_____. 1985b. Effect of depth of house fly pupae in poultry manure on parasitism by six species of Pteromalidae (Hymenoptera). J. Entomol. Sci. 20(4):444-449.

_____. 1985c. Guide to common species of pupal parasites (Hymenoptera: Pteromalidae) of the house fly and other muscoid flies associated with poultry and livestock manure. North Carolina Agric. Res. Ser. Tech. Bull. 278.

_____. 1987. Reproduction of Pteromalidae (Hymenoptera) parasitic on fresh and frozen house fly (*Musca domestica* L.) pupae. Philipp. J. Sci. 116: 313-326.

Rutz, D. A. & R. C. Axtell. 1980. House fly (*Musca domestica*) parasites (Hymenoptera: Pteromalidae) associated with poultry manure in North Carolina. Environ. Entomol. 9:175-180.

Toyoma, G. M. & I. K. Ikeda. 1976. An evaluation of the fly predators at animal farms on leeward and central Oahu. Proc. Hawaii Entomol. Soc. 22:369-379.

Willis, R. R. & R. C. Axtell. 1968. Mite predators of the house fly: A comparison of *Fuscuropoda vegetans* and *Macrocheles muscaedomesticae*. J. Econ. Entomol. 61:1669-1674.

11. Biological Control of Dung-breeding Flies: Pests of Pastured Cattle in the United States

G. T. Fincher

ABSTRACT

The horn fly, *Haematobia irritans* (L.) and face fly, *Musca autumnalis* DeGeer, are important pests of pastured cattle in the United States. Both species reproduce in cattle dung (cowpats) dropped on pastures. Beneficial insects associated with cowpats in the United States are numerous and several species have been shown to play an important role in the regulation of horn fly and face fly populations during certain times of the year. However, these beneficial species still do not adequately prevent horn fly and face fly populations from reaching unacceptable levels during the fly season. Twenty-nine species of natural enemies of dung-breeding flies have been released in the continental United States to aid in the control of horn flies and face flies. Only nine of the twenty-nine species released thus far have been reported to be established but more are expected to be found during the next few years. The impact of the established species of natural enemies on dung-breeding pests has yet to be adequately documented.

INTRODUCTION

The number of cattle in the United States on 1 January 1987 was estimated to be 102,031,000 with a value of $41,491,650,000 (Anonymous 1987). These animals produced approximately 1,000,000,000 cowpats per day with 60-80% of them falling on the surface of pastures. Many species of insects breed in these cowpats including two economically important pests of cattle, the horn fly, *Haematobia irritans* (L.), and the face fly, *Musca autumnalis* DeGeer. These two pest species, neither of which is native, cost livestock producers millions of dollars each year for insecticides, equipment and labor for insecticide applications, and production losses. Many species of the diverse insect fauna found in cowpats have been reported to be beneficial because they parasitize or prey on immature stages of horn flies and face flies, or compete with these flies for the same food source (Harris and Blume 1986). In Missouri, biological control agents plus climatic conditions have been reported to cause greater than 90% mortality to immature horn flies developing in cowpats (Thomas and Morgan 1972).

THE TARGET PESTS

The horn fly is an obligate blood-sucking parasite of cattle that was first reported in this country in 1887 in New Jersey (Riley 1889). Horn flies interfere with normal feeding activity of cattle and cause loss of blood and reduced weight gains. Annual losses in cattle production caused by horn flies are estimated to be in excess of $730 million (Drummond et al. 1981). Most control practices have involved the use of conventional insecticides. However, this pest has developed resistance to many of the insecticides commonly used for its control (Sparks et al. 1985).

Adult female horn flies lay eggs in fresh cowpats, usually within 10 minutes after deposition of the pat. The generation time is eight to thirty days, depending on the temperature (Kunz and Cunningham 1977).

The face fly was detected in Nova Scotia in 1952 (Vockeroth 1953) and quickly spread throughout most of temperate North America (Anonymous 1969). These flies do not feed on blood, but cause annoyance by feeding on the moist mucous secretions of cattle, especially around the eyes and nose. This annoyance causes cattle to cluster and stop feeding. Face flies are also capable of vectoring pinkeye and eye worms (*Thelazia* spp.) of cattle (Moon and Meyer 1985). Annual production losses caused by the face fly are estimated to be in excess of $53 million (Drummond et al. 1981). Female face flies also deposit eggs in fresh cowpats. Within a range of average midsummer dung and soil temperatures (twenty to twenty-six degrees C), face fly development from egg to adult requires twelve to twenty days (Moon and Meyer 1985).

DUNG-INHABITING INSECTS

One of the first ecological studies of insects associated with cattle dung dropped on pasture in the United States was by Mohr (1943); additional surveys of dung-inhabiting insects in various parts of the country have since been reported by Blume (1970), Merritt and Anderson (1977), Poorbaugh et al. (1968), Sanders and Dobson (1966), Valiela (1969), and Wingo et al. (1974). Blume (1985) recently published a checklist, distribution maps, and annotated bibliography of insects associated with cattle dung on pastures in America north of Mexico. His checklist totaled 457 species. In central Texas, 103 species of insects have been found associated with cattle dung dropped on pasture (Blume 1970). Many of these coprophilus insects influence populations of dung-breeding flies. An inverse relationship between the number of horn flies emerging and the number of other insects found in cattle dung has been reported (Blume et al. 1970).

PARASITES

Blume (1985) listed forty-three species of parasitic Hymenoptera associated with cattle dung in the United States. A summary of parasitoids recovered from horn fly and face fly pupae is listed in Table 1. The parasitism rate of the horn fly and face fly is usually low, and only occasionally is the parasitism rate adequate to reduce fly populations. Combs and Hoelscher (1969) in Mississippi and Harris and Summerlin (1984) in Texas reported horn fly parasitism rates as high as 43% and 45%, respectively, with a monthly average of less than 5%. Thomas and Wingo (1968) revealed a seasonal parasitism rate for the face fly in Missouri of less than 9.3%. Also in Missouri, Figg et al. (1983) reported a 3.1% mortality rate of face fly pupae due to parasites.

Legner (1978) reported that *Muscidifurax raptorellus* Kogan and Legner, *M. uniraptor* Kogan and Legner, *M. raptoroides* Kogan and Legner, and *Spalangia* spp. were imported and released in California for filth fly control but that some of them were also adaptable to the horn fly. The three *Muscidifurax* species were reported to be established (Legner 1978); however, it is not known if they parasitize horn fly pupae in the field.

Table 1. Parasitoids Recovered from Horn Fly and Face Fly Pupae in the Continental United States.[a]

Species	Horn Fly	Face Fly
COLEOPTERA		
Staphylinidae		
Aleochara bimaculata Gravenhorst	+	+
Aleochara tristis Gravenhorst	--	+
HYMENOPTERA		
Braconidae		
Alysia sp.	--	+
Aphaereta pallipes (Say)	+	+
Diapriidae		
Trichopria haematobiae (Ashmead)	+	--
Trichopria sp.	+	--
Eucoilidae		
Cothonaspis sp.	+	--
Eucolia impatiens (Say)	+	+
Eucoila sp.	+	+
Kleidotoma sp.	+	--
Pseudeucoila sp.	+	--
Rhoptromeris sp.	+	--
Trybliographa sp.	+	--
Figitidae		
Figites sp.	+	--
Neralsia hyalinipennis (Ashmead)	+	--
Trischiza atricornis (Ashmead)	+	--
Xyalophora quinquelineata (Say)	-	+
Ichneumonidae		
Phygadeuon sp.	+	--

(Continued)

[a] Data based on information published by Blickle (1961), Burton and Turner (1970), Combs and Hoelscher (1969), Figg et al. (1983), Harris and Summerlin (1984), Hayes and Turner (1971), Kessler and Balsbaugh (1972), Klimaszewski (1984), Schreiber and Campbell (1986), Thomas and Wingo (1968), Thomas and Morgan (1972), Turner et al. (1968), Watts and Combs (1977), and Wingo et al. (1974).

Table 1.1 (Cont.)   Parasitoids Recovered from Horn Fly and Face Fly Pupae in the Continental United States.[a]

Species	Horn Fly	Face Fly
Pteromalidae		
Eupteromalus sp.	+	--
Muscidifurax raptor Girault & Sanders	+	--
Muscidifurax sp.	+	--
Spalangia cameroni Perkins	+	--
Spalangia endius Walker	+	--
Spalangia haematobiae Ashmead	+	--
Spalangia nigra Latreille	+	+
Spalangia nigroaenea Curtis	+	--

[a] Data based on information published by Blickle (1961), Burton and Turner (1970), Combs and Hoelscher (1969), Figg et al. (1983), Harris and Summerlin (1984), Hayes and Turner (1971), Kessler and Balsbaugh (1972), Klimaszewski (1984), Schreiber and Campbell (1986), Thomas and Wingo (1968), Thomas and Morgan (1972), Turner et al. (1968), Watts and Combs (1977), and Wingo et al. 1974).

One species of Staphylinidae, *Aleochara tristis* Gravenhorst, an ectoparasite on fly pupae in the larval stage and a predator of Diptera eggs in the adult stage, was introduced from Europe for face fly control (Jones 1967), but apparently had no effect on populations of the pest.

PREDATORS

Predaceous insects have been shown to be important mortality factors of both the horn fly (Thomas and Morgan 1972, Roth et al. 1983, Roth 1989) and face fly (Thomas et al. 1983). Most predators of dung-breeding Diptera are in the families Histeridae, Staphylinidae, and Hydrophilidae. Species associated with cattle dung in the United States are listed in Table 2.

Table 2. Histeridae, Hydrophilidae, and Staphylinidae Associated with Cattle Dung in the Continental United States.

Family/Genus	No. Species	Family/Genus	No. Species
Histeridae		Staphylinidae	
Atholus	1	*Aleochara*	8
Euspilotus	3	*Anotylus*	1
Hister	4	*Atheta*	2
Onthophilus	1	*Amischa*	1
Peranus	1	*Belonuchus*	1
Phelister	3	*Cilea*	1
Saprinus	6	*Falagria*	1
Xerosaprinus	2	*Hyponobrus*	1
Xestipyge	1	*Lathrobium*	1
		Leptacinus	1
		Lithocharis	2
Hydrophilidae		*Oxytelus*	1
Cryptopleurum	2	*Philonthus*	15
Cercyon	10	*Platystethus*	1
Sphaeridium	3	*Quedius*	1
Oosternum	1	*Staphylinus*	2
		Stilicus	1
		Tachinus	1
		Tachyporus	1

Staphylinidae are the most important predators because of diversity and high populations. As larvae and as adults, staphylinids prey on eggs and larvae of dung-breeding Diptera. The genus *Philonthus* contains species which have been shown to be the most effective predators (Thomas and Morgan 1972, Harris and Oliver 1979, Roth 1982, 1989, Roth et al. 1983, Hunter et al. 1989). The histerids have also been shown to be effective predators on horn fly eggs and larvae in laboratory studies (Summerlin et al. 1982a, 1982b). However, the number of histerids occurring in cattle dung in east-central Texas is low (Summerlin et al. 1982a). Histerids are predaceous in both the larval and adult stage. Some Hydrophilidae, such as *Sphaeridium* spp., are also predators of dung-breeding flies. Larvae of *Sphaeridium scarabaeoides* (L.) have been reported to prey on dung-breeding fly larvae including the horn fly (Bourne and Hays 1968). Two exotic staphylinids, *Philonthus flavocinctus* Motschulsky from Southeast Asia and *P. minutus* Boheman from various parts of Africa, Southeast Asia and Australia were obtained from Hawaii and Australia, respectively, and released in Texas in 1987. It is too early to determine if they have become established.

Legner (1986) listed seven exotic species of Histeridae that were released in California for horn fly and face fly control: *Atholus coelestis* Marseul, *Hister caffer* Erichson, *H. chinensis* Quensel, *H. nomas* Erichson, *H. scissifrons* Marseul, *Peranus maindroni* (Lewis), and *Santalus parallelus* (Redtenbacher). It is not known if any of these species have established. Three species of histerids, *Pachylister caffer* Erichson, *Hister nomas* Erichson, and *Atholus rothkirchi* Bickhardt from Africa are being evaluated at the Veterinary Toxicology and Entomology Research Laboratory in College Station, TX for horn fly control and possible release in the near future.

Other predators of dung-breeding flies include a species of Diptera. Larvae of *Ravinia lherminieri* (Robineau-Desvoidy) have been reported to be predaceous on larvae of other dung-breeding flies (Pickens 1981). Also, several species of ants, especially the red imported fire ant *Solenopsis invicta* Buren, prey on immature stages of horn flies (Summerlin et al. 1984). In a study near Ithaca, NY, Schmidtmann (1977) reported that adult face flies and stable flies, *Stomoxys calcitrans* (L.), were the primary prey of vespid wasps (Hymenoptera: Vespidae) and that one adult horn fly was also captured by a wasp. A cricket, *Nemobius fasciatus* (DeGeer), has also been reported to prey on horn fly pupae (Bourne and Nielsson 1967).

COMPETITORS

The principal emphasis for the biological control of dung-breeding flies in pasture habitats in the United States in recent years has been on the use of dung-burying scarabs to reduce breeding habitats (Anderson and Loomis 1978; Blume et al. 1973; Fincher 1981, 1986). Dung-burying scarabs not only compete with the horn fly and face fly for the same food source (cattle dung), they also reduce dung accumulation and improve pastures by increased fertility and improved soil structure (Fincher 1981, Fincher et al.

Table 3. Scarabs Associated with Cattle Dung in the Continental United States.

Genus	No. Species	Genus	No. Species
Aphodius	34	*Geotrupes*	5
Ataenius	12	*Glaphyrocanthon*	1
Ateuchus	3	*Liatongus*	1
Boreocanthon	9	*Melanocanthon*	4
Canthon	7	*Onitis*	1
Copris	5	*Onthophagus*	15
Dichotomius	2	*Phanaeus*	9
Euoniticellus	1	*Pseudocanthon*	1

1981). The largest effort to control dung-breeding Diptera with dung beetles has been in Australia, where a goal of equal or greater importance was pasture improvement (Bornemissza 1960, 1976; Ferrar 1973; Gillard 1967).

Numerous species of scarabs have been reported from cattle dung dropped on pasture in the United States (Table 3). Their effect on fly development is primarily a result of competition and habitat destruction. Although there are many native species of dung-burying scarabs associated with cattle dung in this country, there are very few species that can bury significant amounts of dung from a cowpat within a few days after deposit. Competition and habitat destruction by native beetles is inadequate for controlling dung-breeding pest flies. Therefore, several species of dung beetles with greater potential to rapidly bury cattle dung dropped on pasture have been imported and released as biological control agents for the horn fly and face fly. A negative aspect of dung pat destruction by scarabs is that it could also reduce populations of the predaceous and parasitic dung fauna as well as the targeted pests. Dung burial activity by scarabs has been shown to have negative effects on the reproduction of several species of *Philontus* (Roth 1983).

To date, fifteen exotic species of dung-burying scarabs have been released in several states (Table 4). Five of the fifteen species are known to be established. *Onthophagus gazella* (F.), can be found in the southern tier of states from California to Georgia; *O.*

Table 4. Exotic Species of Dung Beetles Released in Texas to Aid in the Control of the Horn Fly.

Species	Year Released	Country of Origin
Onthophagus gazella (L.)	1972	South Africa via Australia
Euoniticellus intermedius (Reiche)	1979	South Africa via Australia
Onthophagus bonasus (F.)	1980	Pakistan
Onitis alexis Klug	1980	South Africa via Australia
Liatongus militaris (Laporte)	1984	South Africa via Hawaii
Onthophagus taurus Schreber	1985	Europe via Georgia
Onthophagus sagittarius (F.)	1985	Sri Lanka via Hawaii
Gromphas lacordairei (Brulle)	1985	Argentina
Onthophagus binodis Thunberg	1986	South Africa via Hawaii
Onthophagus depressus Harold	1987	South Africa via Georgia
Onthophagus nigriventris d'Orbigny	1987	Kenya via Hawaii
Ontherus sulcator (F.)	1987	Argentina
Copris incertus Say	1987	Mexico via Hawaii
Sisyphus rubrus Paschalidis	1987	South Africa via Australia
Onitis vanderkelleni Lansberge	1987	Kenya via Hawaii

taurus (Schreber), found in Florida in 1975, now occurs in nine southern states and in California; *O. depressus* Harold, established in Florida and Georgia from accidental introductions; *Euoniticellus intermedius* (Reiche), occurs in California and Texas; and *Onitis alexis* Klug, occurs in California. Additional species are expected to become established during the next few years from releases made in Alabama, Arkansas, Georgia, Mississippi, New Jersey, Oklahoma, and Texas.

The effect of these dung-burying beetles on populations of pest flies has yet to be fully evaluated, but a decrease in horn fly populations has been noted on cattle in several states when dung beetle populations are sufficient to bury most cowpats within 24 h after deposition. Roth et al. (1983) reported that scarab competition was a significant mortality factor of the horn fly in east-central Texas. Roth (1989) also reported significant reductions in the numbers of horn flies emerging from individual dung pats due to the dung-burying activity of scarabs.

A NEED FOR ADDITIONAL SPECIES OF NATURAL ENEMIES

Successful biological control of dung-breeding pest flies depends on several ecological factors including seasonal distribution, diel flight activity, and habitat preference of natural enemies. In east-central Texas, horn flies begin activity in late February and are most active from March-November. Parasites of this pest are most active from June-August; predators most active from May-October; and competitors most active from April-October (Fincher et al. 1986, Harris and Summerlin 1984, Hunter 1985, Summerlin et al. 1982a). Because of this "lag time," populations of horn flies increase rapidly in the spring before populations of natural enemies increase accordingly. Additional species of natural enemies are needed, especially early in the fly season, to aid the native fauna in reducing the numbers of horn and face flies developing in cowpats.

REFERENCES CITED

Anderson, J. R., and E. C. Loomis. 1978. Exotic dung beetles in pasture and range land ecosystems. Calif. Agric. 32: 31-32.
Anonymous. 1969. USDA Coop. Econ. Insect Rep. 19: 244-246.
Anonymous. 1987. USDA Agricultural Statistics. U.S. Govt. Printing Office, Washington, DC. 541 p.

Blickle, R. L. 1961. Parasites of the face fly, *Musca autumnalis*, in New Hampshire. J. Econ. Entomol. 54: 802.

Blume, R. R. 1970. Insects associated with bovine droppings in Kerr and Bexar Counties, Texas. J. Econ. Entomol. 63: 1023-1024.

_____. 1985. A checklist, distributional record, and annotated bibliography on the insects associated with bovine droppings on pastures in America north of Mexico. Southwest. Entomol. (Suppl. # 9) 55 p.

Blume, R. R., S. E. Kunz, B. F. Hogan, and J. J. Matter. 1970. Biological and ecological investigations of horn flies in central Texas: Influence of other insects in cattle manure. J. Econ. Entomol. 63: 1121-1123.

Blume, R. R., J. J. Matter, and J. L. Eschle. 1973. *Onthophagus gazella*: Effect on survival of horn flies in the laboratory. Environ. Entomol. 2: 811-813.

Bornemissza, G. F. 1960. Could dung eating insects improve our pasture? J. Aust. Inst. Agric. Sci. 26: 54-56.

_____. 1976. The Australian dung beetle project 1965-1975. Australian Meat Res. Comm. Rev. 30. 32 p.

Bourne, J. R., and R. J. Nielsson. 1967. *Nemobius fasciatus* -- A predator on horn fly pupae. J. Econ. Entomol. 60: 272-274.

Bourne, J. R., and K. L. Hays. 1968. Effects of temperature on predation of horn fly larvae by the larvae of *Sphaeridium scarabaeoides*. J. Econ. Entomol. 61:321-322.

Burton, R. P., and E. C. Turner, Jr. 1970. Mortality in field populations of face fly larvae and pupae. J. Econ. Entomol. 63: 1592-1594.

Combs, R. L., Jr., and C. E. Hoelscher. 1969. Hymenopterous pupal parasitoids found associated with the horn fly in northeast Mississippi. J. Econ. Entomol. 62: 1234-1235.

Drummond, R. O., G. Lambert, H. E. Smalley, Jr., and C. E. Terrill. 1981. Estimated losses of livestock to pests, p. 111-127. In: D. Pimentel (ed.), CRC Handbook of Pest Management in Agriculture, Vol. 1. CRC Press, Inc., Boca Raton, Florida.

Ferrar, P. 1973. The CSIRO dung beetle project. Wool Technol. Sheep Breed. 20:73-75.

Figg, D. E., R. D. Hall, and G. D. Thomas. 1983. Host range and ecolsion success of the parasite *Aphaereta pallipes* (Hymenoptera: Braconidae) among dung-breeding Diptera in central Missouri. Environ. Entomol. 12: 993-995.

148

Fincher, G. T. 1981. The potential value of dung beetles in pasture ecosystems. J. Ga. Entomol. Soc. 16 (Suppl.): 316-333.

———. 1986. Importation, colonization, and release of dung-burying scarabs. p. 69-76. In: R. S. Patterson and D. A. Rutz (eds.), Biological Control of Muscoid Flies. Misc. Pub. Entomol. Soc. Amer., No. 61.

Fincher, G. T., W. G. Monson, and G. W. Burton. 1981. Effects of cattle feces rapidly buried by dung beetles on yield and quality of Coastal bermudagrass. Agron. J. 73: 775-779.

Fincher, G. T., R. R. Blume, J. S. Hunter, III, and K. R. Beerwinkle. 1986. Seasonal distribution and diel flight activity of dung-feeding scarabs in open and wooded pasture in east-central Texas. Southwest. Entomol. Supplement No. 10

Gillard, P. 1967. Coprophagous beetles in pasture ecosystems. J. Aust. Inst. Agric. Sci. 33: 30-34.

Harris, R. L., and L. M. Oliver. 1979. Predation of *Philonthus flavolimbatus* on the horn fly. Environ. Entomol. 8: 259-260.

Harris, R. L., and J. W. Summerlin. 1984. Parasites of horn fly pupae in east central Texas. Southwest. Entomol. 9: 169-173.

Harris, R. L., and R. R. Blume. 1986. Beneficial insects inhabiting bovine droppings in the United States. p. 10-15. In: R. S. Patterson and D. A. Rutz (eds.), Biological Control of Muscoid Flies. Misc. Pub. Entomol. Soc. Amer., No. 61.

Hayes, C. G., and E. C. Turner, Jr. 1971. Field and laboratory evaluation of parasitism of the face fly in Virginia. J. Econ. Entomol. 64: 443-448.

Hunter, J. S., III. 1985. Biology and distribution of staphylinid beetles associated with bovine feces in south-central Texas. Ph.D. dissertation, Texas A&M University

Hunter, J. S., III, D. E. Bay, and G. T. Fincher. 1989. Laboratory and field observations on the life history and habits of *Philonthus cruentatus* and *P. flavolimbatus*. Southwest. Entomol. 14: 41-47.

Jones, C. M. 1967. *Aleochara tristis*, a natural enemy of face fly: Introduction and laboratory rearing. J. Econ. Entomol. 60: 816-817.

Kessler, H., and E. U. Balsbaugh, Jr. 1972. Parasites and predators of the face fly in east-central South Dakota. J. Econ. Entomol. 65: 1636-1638.

Klimaszewski, J. 1984. A revision of the genus *Aleochara* Gravenhorst of America north of Mexico (Coleoptera: Staphylinidae, Aleocharinae). Mem. Entomol. Soc. Can. 129: 211 p.

Kunz, S. E., and J. R. Cunningham. 1977. A population prediction equation with notes on the biology of horn fly in Texas. Southwest. Entomol. 2: 79-87.

Legner, E. F. 1978. Parasites and predators introduced against arthropod pests. Diptera-Muscidae, p. 346-355. In: C. P. Clausen (ed.), Introduced Parasites and Predators of Arthropod Pests and Weeds: A World review. Agric. Handb. No. 480, USDA, U.S. Govt. Printing Office, Wash., DC.

_____. 1986. The requirement for reassessment of interactions among dung beetles, symbovine flies, and natural enemies. p. 120-131. In: R. S. Patterson and D. A. Rutz (eds.) Biological Control of Muscoid Flies. Misc. Pub. Entomol. Soc. Am., No. 61.

Merritt, R. W., and J. R. Anderson. 1977. The effects of different pasture and rangeland ecosystems on the annual dynamics of insects in cattle droppings. Hilgardia 45:31-71.

Mohr, C. O. 1943. Cattle droppings as ecological units. Ecol. Monogr. 13: 276-298.

Moon, R. D., and H. J. Meyer. 1985. Nonbiting flies, p. 5-82. In: R. E. Williams, R. D. Hall, A. B. Broce and P. J. Scholl (eds.), Livestock Entomology. John Wiley & Sons, New York, NY.

Pickens, L. G. 1981. The life history and predatory efficiency of *Ravina lherminieri* (Diptera: Sarcophagidae) on the face fly (Diptera: Muscidae). Can. Entomol. 113: 523-526.

Poorbaugh, J. H., J. R. Anderson, and J. F. Burger. 1968. The insect inhabitants of undisturbed cattle droppings in northern California. Calif. Vector Views 15: 17-36.

Riley, C. V. 1889. The horn fly, *Haematobia serrata* Robineau-Desvoidy. Insect Life 2: 93-103.

Roth, J. P. 1982. Predation on the horn fly, *Haematobia irritans* (L.) by three *Philonthus* species. Southwest. Entomol. 7: 26-30.

_____. 1983. Compatibility of coprophagous scarabs and fimicolous staphylinids as biological control agents of the horn fly, *Haematobia irritans* (L.) (Diptera: Muscidae). Environ. Entomol. 12: 124-127.

_____. 1989. Field mortality of the horn fly on unimproved central Texas pasture. Environ. Entomol. 18: 98-102.

150

Roth, J. P., G. T. Fincher, and J. W. Summerlin. 1983. Competition and predation as mortality factors of the horn fly, *Haematobia irritans* (L.) (Diptera: Muscidae), in a central Texas pasture habitat. Environ. Entomol. 12: 106-109.

Sanders, D. P., and R. C. Dobson. 1966. The insect complex associated with bovine manure in Indiana. Ann. Entomol. Soc. Am. 59: 955-959.

Schmidtmann, E. T. 1977. Muscid fly predation by *Vespula germanica* (Hymenoptera: Vespidae). Environ. Entomol. 6: 107-108.

Schreiber, E. T., and J. B. Campbell. 1986. Parasites of the horn fly in western Nebraska. Southwest. Entomol. 11: 211-215.

Sparks, T. C., S. S. Quisenberry, J. A. Lockwood, R. L. Byford, and R. T. Roush. 1985. Insecticide resistance in the horn fly *Haematobia irritans*. J. Agric. Entomol. 2: 217-233.

Summerlin, J. W., D. E. Bay, and R. L. Harris. 1982a. Seasonal distribution and abundance of Histeridae collected from cattle droppings in south central Texas. Southwest. Entomol. 7: 82-86.

Summerlin, J. W., H. D. Petersen, and R. L. Harris. 1984. Red imported fire ant (Hymenoptera: Formicidae): effects on the horn fly (Diptera: Muscidae) and coprophagous scarabs. Environ. Entomol. 13: 1405-1410.

Summerlin, J. W., D. E. Bay, R. L. Harris, D. J. Russell, and K. C. Stafford III. 1982b. Predation by four species of Histeridae (Coleoptera) on horn fly (Diptera: Muscidae). Ann. Entomol. Soc. Amer. 75: 675-677.

Thomas, G. D., and C. W. Wingo. 1968. Parasites of the face fly and two other species of dung-inhabiting flies in Missouri. J. Econ. Entomol. 61: 147-152.

Thomas, G. D., and C. E. Morgan. 1972. Parasites of the horn fly in Missouri. J. Econ. Entomol. 65: 169-174.

Thomas, G. D., R. D. Hall, C. W. Wingo, D. B. Smith, and C. E. Morgan. 1983. Field mortality studies of the immature stages of the face fly (Diptera: Muscidae) in Missouri. Environ. Entomol. 122: 823-830.

Turner, E. C., Jr., R. P. Burton, and R. R. Gerhardt. 1968. Natural parasitism of dung-breeding Diptera: A comparison between native hosts and an introduced host, the face fly. J. Econ. Entomol. 61: 1012-1015.

Valiela, I. 1969. The arthropod fauna of bovine dung in central New York and sources on its natural history. J. N.Y. Entomol. Soc. 77: 210-220.

Vockeroth, J. R. 1953. *Musca autumnalis* De G. in North America (Diptera: Muscidae). Can. Entomol. 85: 422-423.

Watts, K. J., and R. L. Combs, Jr. 1977. Parasites of *Haematobia irritans* and other flies breeding in bovine feces in northeast Mississippi. Environ. Entomol. 6: 823-826.

Wingo, C. W., G. D. Thomas, G. N. Clark, and C. E. Morgan. 1974. Succession and abundance of insects in pasture manure: Relationship to face fly survival. Ann. Entomol. Soc. Amer. 67: 386-390

12. The Effect of Feeding System on the Dipterous Flies and Predaceous Mites Inhabiting Dung of Farm Animals

E.T.E. Darwish, A. M. Zaki and A. A. Osman

ABSTRACT

The relationships between ration components for livestock and dipterous fly larvae and predaceous mites inhabiting dung of farm animals were studied and discussed. A positive result was detected between wet rations and dipterous fly larvae; an opposite result was observed relative to predaceous soil mites.

Dipterous fly larvae in manure are available as food for predaceous soil mites. Certain species of the families Macrochelidae, Rhodacaridae, Parasitidae, Uropodidae and Cheyletidae are predaceous on fly larvae and play a role as biocontrol agents.

INTRODUCTION

There are large differences in the ability of different animal species to digest bulky feeds. Maynard and Loosli (1969) reported different digestibilities of all nutrients for different animal species. They showed that ruminants digest appreciably more of the nutrients than non-ruminants. The difference is considerably greater for low-quality high fiber feeds. However, when high-energy diets are fed, non-ruminants exhibit digestive ability equal to the ruminants. Even within ruminants, the different species have different digestive ability (Abdel-Rahman and Ahmed, 1984). Therefore, it is expected to have different kinds of manure not only with different animal species, but also with different kinds of rations.

The habits of the flies are quite variable. They can be predators, parasites, scavengers or attack man and domestic animals. The actions of dipterous larvae in manure produce an unsightly condition, increase the problem of unpleasant odors, and sometimes cause spread of manure onto the walkways. During severe outbreaks, biting flies can cause nervous animals, reduce the weight of beef, and reduce the production of milk from 10% to 50% (Davidson, 1966 S Miller et al., 1973; Campbell, 1976; Arends et al., 1982 and Kunz et al., 1984).

Domestic animal manure is suitable habitat for a very large cosmopolitan group of mites which have been recorded as predators and scavengers. Many predaceous soil species are considered beneficial to man since they feed on harmful arthropods (Hurbutt, 1958; Axtell, 1961; 1963 a & b; 1968, 1969 and Berry, 1973). Members of the family Macrochelidae have been utilized successfully in house fly control in manure. For example, *Macrocheles muscaedomesticae* (Scop.) achieved a 99% kill of house fly eggs in poultry manure (Rodriguez et al., 1970), substantially reducing the house fly population. *Fuscuropoda vegetans* (DeGeer) (Gamasida, Uropodidae) is a common predator of early instar fly larvae (O'Donnell and Axtell, 1965), as is *Parasitus coleoptratorum* (L.) (Parasitidae) (Wernz and Krantz, 1976).

The feeding systems of four farm animals (buffalo,cattle, sheep and goat) were examined to determine if there are relationships between rations and manure components and their effect on the seasonal variations of dipterous flies and predaceous mites.

MATERIALS AND METHODS

The animals under investigation (buffalo, cattle, sheep and goat) at the Experimental Farm Station of Faculty of Agriculture, Menoufia University, El-Rahib, Egypt were fed the rations shown in Table 1 during one successive year.

Four dung samples of the farm animals were collected by the aid of an iron sampler (10 X 10 X 10 cm); replicated three times and transported immediately for extraction by modified Tullgren funnels into 70% ethyl alcohol for 72 hrs. Counts were made under a dissecting microscope for dipterous individuals, while mite specimens were collected and identified using microscopic examination with the so-called half-open slide with lactic acid. Permanent slides for rapid comparison purposes were made using Berless's fluid. Thereafter, the mites were identified according to Krantz (1978) and Balogh and Mahunka (1983).

Table 1. **Feeding system and chemical composition of the rations fed to animals during the year.**

Mo.	Ration	CP	EE	NFE	CF	A	Total
			%	Dry	Matter		
Dec.	Wheat straw, corn stalk, co-op concentrates, corn grain (with cobs) and clover.	7.3	2.1	39.0	20.2	4.4	73.0
Jan.	Corn stalk, barley, wheat bran, cotton seed meal and clover.	11.7	2.0	27.6	10.3	4.0	55.6
Feb.	Corn stalk, co-op concentrates, wheat bran, barley, clover and fodder beet.	6.0	1.8	23.0	7.4	3.8	42.0
Mar.	Corn stalk, co-op concentrates, wheat bran, barley, clover and fodder beet.	6.0	1.8	23.0	7.4	3.8	42.0
Apr.	Wheat bran, corn stalk, clover and clover hay.	7.3	1.9	26.7	14.4	4.6	54.9
May	Wheat bran, corn stalk, clover, clover hay and supplemented with co-op concentrates.	8.6	2.1	30.6	15.2	5.2	61.7
June	Corn stalk, co-op concentrates, wheat bran, barley and fodder beet.	9.4	1.9	43.7	12.8	5.6	73.4
July	Wheat straw, corn stalk, barley, wheat bran and co-op concentrates.	9.6	2.2	48.8	20.6	6.1	87.3

* CP = Crude protein, EE = Ether extract, NFE = nitrogen free extract, CF = Crude fiber and A = Ash.

(Continued)

Table 1.1 (Cont.) Feeding system and chemical composition of the rations fed to animals during the year.

		Chemical Composition % Dry Matter					
Mo.	Ration	CP	EE	NFE	CF	A	Total
Aug.	Wheat straw, corn fodder (darawa), fodder beet and co-op concentrates.	7.3	1.8	34.1	19.8	6.7	69.7
Sept.	Wheat straw, corn fodder, fodder beet, cotton seed meal and co-op concentrates.	15.0	1.9	31.8	18.5	6.7	73.9
Oct.	Wheat straw, wheat bran, co-op concentrates and cotton seed meal.	13.7	2.4	46.4	20.0	7.0	89.5
Nov.	Wheat straw, corn stalk, wheat bran, cotton seed meal and co-op concentrates.	12.5	2.1	42.4	23.4	7.0	87.4

* CP = Crude protein, EE = Ether extract, NFE = nitrogen free extract, CF = Crude fiber and A = Ash .

RESULTS AND DISCUSSION

Monthly investigations were carried out during one successive year from December to November. Results are presented in Tables 2, 3, and 4 and illustrated by Figures 1 and 2. The monthly variation in the population density of dipterous flies and mites inhabiting dung of four farm animals was calculated as follows:

$$\text{Monthly variation} = \frac{\text{Count given at month}}{\text{Count in a preceding month}}$$

These variation values provide an indication of the favorable periods during the year.

Seasonal variations of dipterous flies

Twenty four species of dipterous flies, belonging to eighteen genera and twelve families were collected during the one year study from the dung of buffalo, cattle, sheep and goat.

From results in Tables 2, 3 and illustration of Figure 1, it is obvious that dipterous fly populations in four kinds of animal dung, vary considerably from one season to another. It appears that the dipterous species have a maximum peak during April, May and June, then decrease gradually reaching their minimum during autumn. However, in winter dipterous flies have low populations, which could be attributed to unfavorable environmental conditions. Because of favorable environmental factors, in addition to the percentages of dung moistures from 59.0%, to 74.7% in the spring, it is easily understood why the dipterous species have the highest peak of seasonal abundance during this period when the monthly variation was 3.4. In summer, dipterous fly populations were moderate in July and August, when the percentages of manure moistures ranged between 47.8% and 59.2%. A low population of flies was observed in the autumn (from September to November) during which time the manure moisture percentages ranged from 46.6% to 55.6%. The highest percentage of relative abundance for dipterous individuals was calculated from buffalo dung (32.1%), followed by cattle dung (29.1%), sheep dung (26.1%) and the lowest proportion was in goat dung (12.6%).

Data in Table 2, reveal that buffalo and cattle manures had two peaks of dipterous individuals; the highest during April at moisture percentage 67.6% and 69.6%, respectively, the second which was considerably lower during November at moisture percentage 47.5% and 50.0%, respectively. In sheep dung, dipterous flies had three annual peaks during January, May and August, when manure's moisture percentages were 54.0%, 59.0% and 47.7%, respectively. While goat dung had a peak during June at a moisture percentage of 54.9%. The favorable period for dipterous species in the four kinds of dung under consideration was observed during March, where the monthly variation values were 2.9, 3.0, 4.4 and 6.0 for buffalo, cattle, sheep and goat, respectively, as manure moisture percentages ranged between 69.4% and 74.7%.

Table 2. Effect of manure moisture on the occurrence of dipterous flies and predaceous mites inhabiting dung of four farm animals.

Month	Buffalo			Cattle			Sheep			Goat		
	Moisture %	Flies No.	Mites No.	Moisture %	Flies No.	Mites No.	Moisture %	Flies No.	Mites No.	Moisture %	Flies No.	Mites No.
Dec.	56.2	35	0	55.1	27	16	50.0	26	83	53.8	11	68
Jan.	66.6	38	0	66.6	24	4	54.0	44	47	56.0	11	48
Feb.	74.8	35	0	73.6	27	6	70.0	16	26	71.1	5	20
Mar.	69.9	100	46	74.7	82	10	69.4	70	49	71.0	30	24
Apr.	67.6	125	16	69.6	121	43	64.0	81	64	66.8	45	51
May	63.0	110	7	63.9	114	13	59.0	103	28	59.7	65	24
June	56.2	98	6	59.2	85	0	53.4	97	12	54.9	71	19
July	47.8	76	15	51.3	71	3	49.3	51	2	50.8	22	3
Aug.	58.6	72	24	51.0	67	33	47.7	85	74	49.7	28	11
Sep.	55.6	36	22	54.0	33	13	50.7	25	0	53.4	14	222
Oct.	46.6	40	15	51.0	42	50	48.6	25	179	53.3	10	81
Nov.	47.5	49	23	50.0	45	41	47.5	37	84	49.8	8	128

Table 3. Total numbers and monthly variation of dipterous flies and mites inhabiting dung of four farm animals during a year.

Month	% Moisture in ration	Dipterous flies		Manure mites		Predacious mites		Saprophagous mites	
		Total	M.V.*	Total	M.V.*	Total	M.V.*	Total	M.V.*
December	27.0	99		426		167		259	
January	44.4	117	1.2	255	0.6	99	0.6	156	0.6
February	58.0	83	0.7	77	0.3	52	0.6	25	0.2
March	58.0	282	3.4	238	3.1	129	2.5	109	4.4
April	45.1	372	1.3	339	1.4	174	0.6	165	1.5
May	38.3	392	1.1	140	0.4	72	0.4	68	0.4
June	26.5	351	0.9	56	0.4	37	0.5	19	0.3
July	12.7	220	0.6	104	1.9	23	0.6	81	4.3
August	30.3	252	1.1	377	3.6	142	6.2	235	2.9
September	26.1	108	0.4	652	1.7	257	1.8	395	1.7
October	10.5	117	1.1	784	1.2	325	1.3	459	1.2
November	12.6	139	1.2	767	1.0	276	0.8	491	1.1

*M.V. = Monthly Variation

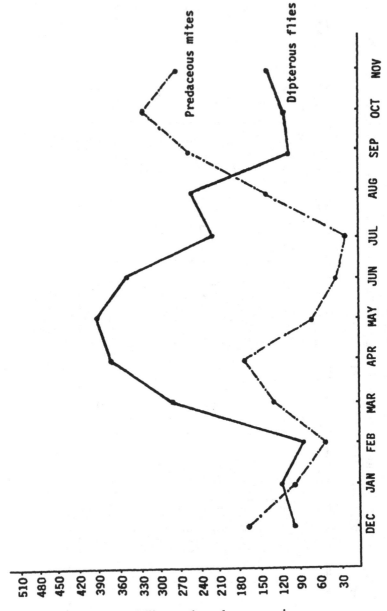

Figure 1. Seasonal abundance of dipterous flies and predaceous mites inhabiting dung of four farm animals.

Seasonal variations of mites

Seasonal abundance of total mites (predaceous and saprophagous) inhabiting dung of the four tested animals occurred with two peaks in buffalo, cattle and goat manures in spring and autumn, while in sheep manure mites have three peaks during spring, summer and autumn (Table 2).

Data in Table 3 show that the population density of total mites had two favorable periods in March and August, where the monthly variation values were 3.1 and 3.6. The favorable periods for predaceous mite activity were also in March with monthly variation of 2.5 and in August with a maximum monthly variation value 6.2. Sharp declines in the population of predaceous mites were observed during winter and early summer.

On the other hand, the highest numbers of predaceous mites observed during a whole year were in goat manure, followed by sheep manure, buffalo manure and cattle manure (Figure 2). With saprophagous mites, data in Table 3 reveal that seasonal variations have two favorable periods during spring and summer, when the monthly variation values were 4.4 and 4.3 in March and July, respectively.

Mite species composition

Table 4 shows that the present investigations revealed twenty families representing twenty-five genera in the suborders Gamasida, Actinedida, Acaridida and Oribatida. Gamasid mites represented 81.6%, 55.2%, 68.2% and 63.8% of total predaceous mites in the manure of buffalo, cattle, sheep and goat, respectively. The macrochelid mites were the most dominant and frequent species in all dung kinds.

Filipponi (1960) mentioned that *Macrocheles muscaedomesticae* (Scop.) was especially common as a phoretic on *Musca domestica* Macp., *Stomoxys calcitrans* (L.), and *Fannia canicularis* (L.). Petrova (1964) reported that *M. muscaedomesticae* had been recovered from eleven fly species. Decreases of 83% to 92% in populations of house fly and *Fannia canicularis* (L.) were recorded by Axtell (1963 a & c) and by Singh et al., (1966) in experiments where mite populations were augmented. Axtell (1969) listed eleven behavioral, biological and structural characteristics of macrochelid mites which he considered to be assets in the biological control of dung-breeding flies.

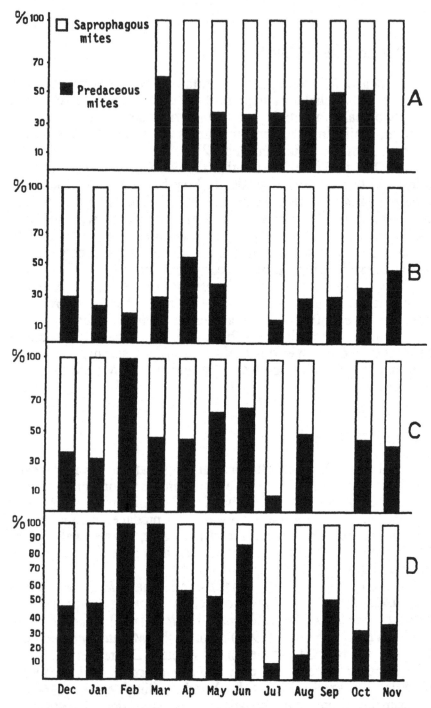

Figure 2. Relative abundance of predaceous mites in manure of A, buffalo; B, cattle; C, sheep and D, goat.

Table 4. Mite species associated with dung of four farm animals in Egypt.

Suborders of Acari	Species	Animal dung/ associates	Frequency
Suborder: Gamasida			
Superfamily: Parasitoidea			
Family: Parasitidae	*Parasitus* sp.	All animals	**
	Parasitus coleoptratorum	All *animals*	**
	Poecilochirus sp.	All animals	*
Superfam: Rhodacaroidea			
Family:Rhodacaridae	*Rhodacarus* sp.	All animals	*
Family: Digamasellidae	*Digamasellus* sp.	All animals	**
Family: Ologamasidae	*Gamasellus* spp.	All animals	**
Superfam: Eviphidoides			
Family: Macrochelidae	*Macrocheles* sp.	All animals	***
	M. muscae-domesticae	All animals	***
Superfam: Dermanyssoidea			
Family: Laelapidae	*Androlaelaps* sp.	Sheep & Buffalo	*
Superfam.: Uropodoidea			
Family: Uropodidae	*Fuscuropoda* spp.	Sheep & Buffalo	**
Suborder: Actinedida			
Superfam: Pyemotoidea			
Family: Pyemotidae	*Pyemotes* sp.	Sheep	*
Superfam:Pygmephoroidea			
Family:Pygmephoridae	*Pediculaster* sp.	Sheep & Goat	**
	P. mesembrinae	Sheep & Goat	**
	Bakerdania n tarsalis	Sheep & Goat	**
	B. centriger	Sheep & Goat	**
	B. exigua	Sheep & Goat	**
Family: Scutacaridae	*Pymodispus* sp.	Goat	*
	Scutacarus eucomus	Goat	*

* Occasionally, ** Frequently and *** Always.   (Continued)

Table 4.1 (Cont.) Mite species associated with dung of four farm animals in Egypt.

Suborders of Acari	Species	Animal dung/ associates	Frequency
Superfam: Cheyletoidea			
Family:Cheyletidae	*Cheyletus* sp.	Sheep & Goat	***
Superfam: Erythraeoidea			
Family: Erythaeidae	*Balaustium* sp.	Sheep & Goat & Cattle	***
Suborder: Acaridida			
Superfam.: Acaroidea			
Family: Glycyphagidae	*Glycypyhagus* sp.	All animals	***
Family: Acaridae	*Tyrophagus* sp.	All animals	**
	Rizoglyphus sp.	All animals	**
Family: Saprohlyphidae	*Saproglyphus* sp.	All animals	**
Superfam: Anoetoidea			
Family: Anotidae	*Anoetus* sp.	Sheep & Goat	***
	Histiostoma sp.	Sheep & Goat	***
Suborder: Oribatida			
Superfam.: Oribatulodea			
Family: Oribatulidae	*Achipteria* sp.	All animals	**
Superfam:Ceratozetoidea			
Family: Ceratozetidae	*Trichoribates* sp.	All animals	**
Superfam: Hermannielloidea			
Family: Hermanniellidae	*Hermanniella* sp.	All animals	**
Superfam: Liodoidea			
Family: Liodidae	*Platyliodes* sp.	All animals	*

* Occasionally, ** Frequently and *** Always.

Predaceous mites of suborder Actinedida represented 13.4%, 9.0%, 8.6% and 4.0% of the total predaceous mites in the dung of cattle, sheep, goat and buffalo, respectively. Individuals of the families Cheyletidae and Erythraeidae were the most dominant in sheep and goat manures.

Some acaridid mites are known as necrophaga on dead soil insects. Another acaridid acarid, *Rizoglyphus echinopus* F. et R. may be a macrophytophaga (Hussey et al., 1969). Members of the acaridid family Anoetidae have highly modified mouthparts which allow them to filter microorganisms from liquified substrates

(Hughes and Jackson, 1958). In the present work anoetid mites represented 31.4%, 27.6%, 22.8% and 14.4% in the dung of cattle, goat, sheep and buffalo, respectively

Oribatid mites are typical members of the edaphic community that take advantage of the comparatively rich dung habitat when it becomes available to them. Luxton (1972) considered oribatid mites as a saprophagous group.

From the above mentioned data, a relationship was observed between some predaceous mite species and dipterous fly species inhabiting dung of four kinds of farm animals, in which the population of dipterous flies was low and the population of mites was high. This relationship may be due to certain mite species of the families Macrochelidae, Rhodacaridae, Parasitidae, Uropodidae and Cheyletidae which are predaceous on fly larvae and play a role as biocontrol agents. Axtell (1963 a and c) reported seven species of Macrochelidae mites; four were common in manure from dairy cattle and he also reported that the reduction in numbers of the house fly, *Musca domestica*, by the mites was large compared to the actual fly production from manure.

This study indicated that the feeding systems of four farm animals may be an indirect factor in the spread of dipterous flies and mites (predaceous and saprophagous) throughout the four seasons of the year. In this respect, the population density of dipterous species increased by increasing the moisture in the manure of the previous animals which was due to the high moisture in the ration components during the spring season. On the other hand, the population density of predaceous mites had a negative response to wet rations, with the number of predaceous mites in the manures of the tested farm animals decreasing with increasing manure moisture.

There was a negative relationship between moisture and crude protein percentages in the rations. Rations containing high moisture and low protein lead to an increase in fly numbers and a decrease in predaceous mites. On the contrary, rations having low moisture and high protein percentages lead to a relative decrease in fly numbers and an increase in predaceous mites.

The present results are supportive of the findings of Axtell (1963c, 1969 and 1981) who noted that an integrated approach using mite and insect predators, parasites, selective use of insecticides and baits, and culture practices that ensure low moisture content of the dung substrate would be an appropriate strategy for fly control in domestic settings.

166

ACKNOWLEDGMENT

The authors acknowledge Dr. Barakat M. Ahmed, Associate Professor of Animal Nutrition, Dept. of Animal Production, Menoufia Univ., for his cooperation and making the needed facilities available and for the analysis of these materials.

REFERENCES CITED

Abdel-Rahman, K. and B. M. Ahmed. 1984. Allantoin in urine by ruminant in relation to energy intake. Proc. 1st Egypt-British Conf. on Anim. and Poultry Prod. Zagazig, Sep 11-13:376-381.

Arend, J. J.; R. E. Wright; K. S. Lusby and R W. McNew. 1982. Effect of face flies (Diptera: Muscidae) on weight gains and feed efficiency in beef heifers. J. Econ. Ent 75(5):794-797 .

Axtell, R. C. 1961. New records of North American Macrochelidae (Acarina:Mesostigmata) and their predation rates on the house fly. Ann. Ent. Soc. Amer., 54 (5):748.

_____. 1963a. Effect of Macrochelidae Acarina: Mesostigmata) on house fly production from dairy cattle manure. J. Econ Ent. 56(3):317-321.

_____. 1963b. Acarina occurring in domestic animal manure. Ann. Ent Soc. Amer., 56(5):628-633 .

_____. 1963c. Manure-inhabiting Macrochelidae (Acarina: Mesostigmata) predaceous on the house fly. Adv. Acarol. 1:55-59.

_____. 1968. Integrated fly control: Populations of fly larvae and predaceous mites, *Macrocheles muscaedomesticae*, in poultry manure after larvicide treatment. J. Econ. Ent. 61(1):245-249.

_____. 1969. Macrochelidae (Acarina:Mesostigmata) as biological control agent for synanthropic flies. Proc. 2nd Int Cong. Acarol. (1967):401-405.

_____. 1981. Use of predators and parasites in filth fly IPM programs in poultry housing. Page 26-43 in Status of Biological control of filth flies. Workshop Proc. USDA A 106 2:F 64 212 pp. C f Proc. Conf. of Biological Control of.Pests by mites. Calif. Univ., Berkeley April 5-7, 1982:185 pp.

Balogh, J. and S. Mahunka. 1983. The soil mites of the world. I-Primitive oribatids of the palaearctic region. Akad. Kiadb, Budapest, 372 pp.

Berry, R. E. 1973. Biology of the predaceous mite, *Pergamasus quisquiliarum* on the garden symphylan *Scutigerella immaculata*, in the laboratory. Ann. Ent. Soc. Amer. 66(6):1354-1356. C.F. Krantz, 1978: A manual of acarology. 509 pp.

Campbell, J. B. 1976. Effect of horn fly control on cows as expressed by increased weaning weights of calves. J. Econ. Ent. 69(6):711-712.

Davidson, H. 1966. Insect pest of farm, garden, and orchard. Sixth edition. John Wiley & Sons, New York, London, Sydney, Toppan Company, Ltd., Tokyo, Japan. 675 pp.

Filipponi, A. 1960. Macrochelidi (Acarina:Mesostigmata) foreticidi mosche risultan parziati di una indagine ecologica n corso nell'Agro Pontino Parassitol. Z. (1-2):167-172. C.f. Proc. Conf. of Biological Control of Pests by mites. Calif. Univ., Berkeley. April 5-7, 1982. 185 pp.

Hughes, R. D. and C. G. Jackson. (1958). A review of the Anoetidae (Acari). Virginia J. Sci. 9 N.S.(1):5-198. C.f. Krantz, 1978: A manual of acarology. 509 pp.

Hurlbutt, H. W. 1958. A study of soil inhabiting mites from Connecticut apple orchards. J. Econ. Ent. 51(6):76-82.

Hussey, N.W., W.H. Read and J.J. Hesling. (1969). Order Acarina: Mites In the pests of protected cultivation. Amer. Elsevier Publ. Co., New York: 190-228. C.f. Krantz, 1978: A manual of acarology, 509 pp.

Krantz, G. W. 1978. A manual of acarology. Second Edition Publ. by Oregon State Univ. Book Stores, INC. 509 pp.

Kunz, S. E., J. A. Miller, P. L. Sims and D. C. Meyerhoeffer. 1984. Economics of controlling horn flies (Diptera:Muscidae) in range cattle management. J. Econ. Ent. 77(3):567-660.

Luxton, M. 1972. Studies on the oribatid mites of a Danish beech wood soil.Pedobiol. 12:434-463. C.f. Krantz, 1978. A manual of acarology. 509 pp.

Maynard, L. A. and J. K. Loosli. 1969. Animal Nutrition. 6th ed. TATA McGraw-Hill Publ. Co., New York. 248 pp.

Miller, R. W., L. G. Pickens, N. O. Morgan, R. W. Thimijan and R. L. Wilson. 1973. Effect of stable flies on feed intake and milk production of dairy cows. J. Econ. Ent. 66(9):711-713.

O'Donnell, A. E. and R. C. Axtell. 1965. Predation by *Fuscuropoda vegetans* (Acarina:Uropodidae) on the house fly (*Musca domestica*). Ann. Ent. Soc. Amer. 58:403-404. C.f. Krantz, 1978. A manual of acarology. 509 pp.

Petrova, A.D. 1964. The part played by synanthroplic flies in the spread of Macrochelidae Vitzt. Med. Parazit. (Moscow) 33(5):553-557 (in Russian). C.f. Proc. Conf. of Biological Control of Pests by Mites. Calif. Univ., Berkeley. April 5-7, 1982:185 pp.

Rodriguez, J. G., P. Singh and B. Taylor. 1970. Manure mites and their role in fly control. J. Med. Ent. 7(3):335-341. C.f. Krantz, 1978. A manual of acarology. 509 pp.

Singh, P., W. King and J. G. Rodriguez. 1966. Biological control of muscids as influenced by host preference of *Macrocheles muscaedomesticae* (Acarina:Macrochelidae). J. Med. Ent. 3:78-81. C. f. Proc. Conf. of Biological Control of Pests by Mites. Calif. Univ., Berkeley. April 5-7, 1982. 185 pp.

Wernz, J. G. and G. W. Krantz. 1976. Studies on the function of the tritosternum in selected Gamasida (Acari). Can. J. Zool. 54:202-213. C.f. Krantz, 1978. A manual of acarology. 509 pp.

13. *Hydrotaea (Ophyra)* Species as Potential Biocontrol Agents Against *Musca domestica* (Diptera) in Hungary

Róbert Farkas and László Papp

H. aenescens and *H. ignava* are bright bluish-black or brassy-black garbage flies or dump flies, better known as *Ophyra* species. Because they are so closely related to the species *Hydrotaea*, their genus has been amalgamated with *Hydrotaea* (a very recent reference work on this problem is Pont's (1986) contribution in the *Catalogue of Palaearctic Diptera*.) The name of *Ophyra leucostoma* has changed twice in the last few years. The specific name *ignava* (Harris 1780) was proved to be a senior synonym of *leucostoma* (Wiedemann 1817).

As early as 1923, Séguy wrote about the carnivorous habits of *Ophyra leucostoma*. His findings were confirmed by Keilin and Tate (1930), who found a parallelism between the structure of the cephalopharyngeal skeleton and its carnivorous habits, although it is saprophagous until the 2nd larval instar (Skidmore 1973). It is known that they can fully develop in fresh carrion without preying on living larvae. Derbeneva-Ukhova (1940) regarded *O. leucostoma* larvae as useful predators against larvae of other species of filth flies, as did Leikina (1942) for the larvae of *O. capensis*. Hennig (1964) and Skidmore (1973) are good sources for the old literature concerning their morphology, taxonomy and ecology.

However, until the seventies *Ophyra* species were not regarded as true biocontrol agents against house flies or other filth flies. In the last decade two series of papers have been published on *Ophyra aenescens*, one of the series in the GFR (El-Dessouki and Stein 1978; Stein et al. 1977), the other in the GDR (Bauermeister and Schumann 1980; Schumann 1982; Müller et al. 1981; Müller 1982). The latter group stressed their importance in biological control, e.g., Schumann (1982) says: "The species (*O. aenescens*) is probably a suitable antagonist of *Musca domestica*." However, some American authors do not agree with their optimism (Axtell and Rutz 1986).

In the course of our project to map the breeding sites of house fly larvae for the purpose of more effective fly control, *Hydrotaea leucostoma* (*ignava*) larvae were found to be aggressive killers of house fly larvae in the manure of pig sties, as well as in the litter of caged-layer houses (see also Anderson and Poorbaugh 1964).

The other part of our work on *Ophyra* species was to collect more data on *H. aenescens* as a potential biocontrol agent against *M. domestica* and to adopt a correct view of its control potential, since we felt that the data published by German authors are not convincing in some respects.

MATERIALS AND METHODS

Litter samples (200 g each) were collected from the breeding sites of fly larvae in a caged-layer house. Developed imagos were later identified and counted.

The design of our experiments with *H. aenescens* and *M. domestica* larvae was much influenced by Müller's (1982) work but contrary to his experiments, pig manure -- that is the most natural material for house flies -- was used as a larval medium. 100 g each of manure were put in small plastic cups, inoculated with larvae and put in a bigger glass jar with ca 2 cm high wet sawdust for fly pupation. Cultures (three or more replicates for each trial) were kept in the laboratory at 25± 3 degrees C, and wetted if needed. Imagos and pupal shells were then counted.

RESULTS

The results of the litter samples from the caged-layer house are presented in Table 1. Contrary to our observations by naked eye on the larval mix of *M. domestica* and *H. ignava* (see also Anderson and Poorbaugh 1964), where *Hydrotaea* larvae killed house fly larvae, a chi-square-test and one of its modifications (Yates) indicated only a slight negative correlation but this was not significant. Because interspecific relations are more complex (not mentioning that simple presence-absence relations are too crude to describe real phenomena), the number of imagos developing cannot

Table 1. *M. domestica* and *H. ignava* Developed from Litter Samples of a Caged-Layer House (Number of Samples).

H. ignava

	+	-	Σ
+	2	2	4
-	6	2	8
Σ	8	4	12

H. ignava

	+	-	Σ
+	3	5	8
-	7	11	18
Σ	10	16	26

be evaluated by these simple methods. That is, field experiments are not accurate for a judgment of these relations. A schematic model (Table 2) was set up for analyzing (or rather illustrating) the events in the larval stages. The killing effect of *H. ignava* larvae was not obviously detectable against *Fannia canicularis*, the second most important fly in caged-layer houses in our country (see a forthcoming paper).

In the second part of our study it was observed that if first instar larvae of both species were put together in equal ratios (Table 3), the mortality rate of house fly larvae was a little higher and that of *H. aenescens* was somewhat lower than values when reared in unmixed colony cultures (cf. Müller 1982). The mortality of *M. domestica* increased to almost 100% if its first instars were reared together with second instars of *H. aenescens*. A complete (100%) mortality was observed when first, second and third instar house fly larvae were put into the culture with third instars of *H. aenescens*. Based on our earlier observations and on the results of Müller (1982), we believe that the amount of rearing medium was not a limiting factor in our experiments.

It can be seen in Table 4 that *H. aenescens* larvae also killed all or nearly all house fly larvae in a ratio of three to one for *M. domestica*.

Table 2. Possible Events in Larval Stages.

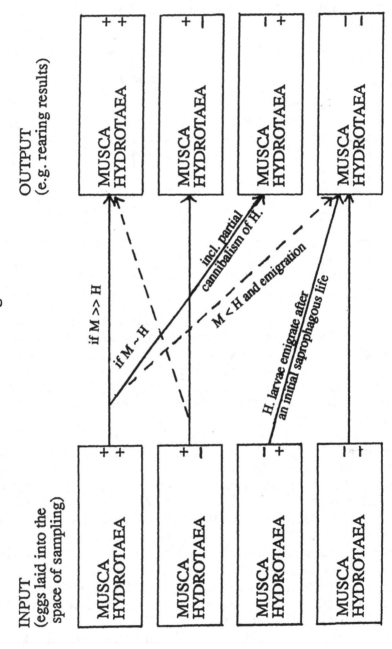

Table 3. Larval Mortality in Cultures of Different Larval Instars of *H. aenescens* and *M. domestica.*

| | | Hydrotaea aenescens | | |
		L1	L2	L3
	L1	28,3xx 25,0xxx	98,3 26,7	100 16,7
Musca				
domestica	L2	----	----	100 23,3
	L3	no effect	----	100 18,0

x   20 ex./species each in 3 repl.
xx mortality rate of *M. domestica* in %
xxx mortality rate of *H. aenescens* in %

Table 4. Larval Mortality in Different Initial Ratios.

| | | Hydrotaea aenescens | | |
| | | L1 | L2 | L3 |
	ex.	20	20	10
	20	28,3xx 25,5xxx	100 16,7	96,7 13,3
Musca domestica				98,9 (------------)
L1	30	----	----	26,7

CONCLUSIONS

Opinions on *Hydrotaea* species as possible biological agents against house flies are rather conflicting. Moon and Meyer (1985) stated that *Ophyra* species are not suitable candidates for biological control of house flies because the adults can be annoying if abundant; however, the adults do not habitually enter buildings. Axtell (1985) shares this view: "*Ophyra* larvae actively prey upon other fly larvae and, in that sense, are beneficial. Overall, however, *Ophyra* are pests because often large numbers are produced in poultry houses and disperse readily." To the contrary, El-Dessouki and Stein (1977) found that *O. aenescens* did not show any great tendency to disperse and was recaptured within a small range (less than 50 m). Axtell and Rutz (1986) summarized the opinions of the American authors: "The larvae of *Ophyra* will prey upon other fly larvae. However, the adults of *Ophyra* are pestiferous and the simple fact of their larvae being capable of predation on other fly larvae does not qualify them as practical biological control agents." Please note the American researchers decribed their possible role in poultry houses.

On the contrary, in the GDR a patent has been applied for mass rearing and spreading of *H. aenescens* in whole-herd pig confinement systems, mainly in deep-pit houses of young pigs where larvae develop in moist manure and other fly control methods are extremely difficult. Satisfactory, or in some cases excellent, fly control was achieved with *H. aenescens* in large pig farms in the GDR. (A fly control symposium was held in Cottbus in 1982, almost exclusively for *H. aenescens* and additional research has been carried out since then for an industrial application.)

Our aim was not to be an umpire between the two schools of thought and our results do not seem to be decisive. Our data only confirm that the larval development of *M. domestica* can be suppressed or prevented by the presence of *H. aenescens* or *H. ignava*. This effect of *Hydrotaea* species are more evident if their larvae are older than the larvae of *M. domestica*. We would like to continue with experiments of different initial age and ratios of *M. domestica* and *H. aenescens* larvae, and to try laboratory rearing of *H. ignava*, in order to obtain reliable data on their potential as biological control agents.

We do not believe that *Hydrotaea* species, as other biological control agents, can be used alone against house fly or other filth breeding flies in livestock and poultry production facilities.

However, they may offer the possibility for use in integrated fly management programs.

REFERENCES CITED

Anderson, J. R., and J. H. Poorbaugh. 1964. Biological control possibility for house flies. California Agriculture 18(9):2-4.
Axtell, R. C. 1985. Arthropod pests of poultry, pp. 269-295. In: Williams, R. E., R. D. Hall, A. B. Broce, and P. J. School (eds.), Livestock Entomology. John Wiley and Sons, New York, 335 p.
Axtell, R. C., and D. A. Rutz. 1986. Role of parasites and predators as biological fly control agents in poultry production facilities, pp. 88-100. In: Patterson, R. S., and D. A. Rutz (eds.), Biological Control of Muscoid Flies. Misc. Publ. ESA, 61:1-174.
Bauermeister, C.-D., and H. Schumann. 1980. Ophyra aenescens (Wied.) - eine für die DDR neue Muscidenart (Diptera). Faun. Abh. Mus. Tierk. Dresden 7(23):213-217.
Derbeneva-Ukhova, V. P. 1940. K ekologii navoznykh mukh v Kabarde. [On the ecology of the filth flies in Kabarda, in Russian.] Medskaya Parazit. 9(4):323-339.
El-Dessouki, S., and W. Stein. 1978. Untersuchungen über die Insektenfauna von Mülldeponien III. Die Ausbreitungstendenz von Fliegen einer Rottedeponie (Dipt., Muscidae und Calliphoridae). Z. angew. Zool. 65/3/:367-375.
Hennig, W. 1964. 63b. Muscidae. In: Lindner, E. (ed.), Die Fliegen der palaerktischen Region, 7/3, Stuttgart, 1110 p.
Keilin, D., and P. Tate. 1930. On certain semi-carnivorous Anthomyid larvae. Parasitology 22:161-181.
Leikina, L.I. 1942. The role of various substrata in the breeding of Musca domestica. [In Russian.] Medskaya Parazit. 11:82-86.
Moon, R. D., and H. J. Meyer. 1985. Nonbiting flies, pp. 65-82. In: Williams, R. E., R. D. Hall, A. B. Broce, and P. J. School (eds.), Livestock Entomology. John Wiley & Sons, New York, 335 p.
Müller, P. 1982. Zur Bedeutung des Musca domestica-Antagonisten Ophyra aenescens (Diptera: Muscidae). III. Laborversuche zur Wechselwirkung zwischen den Larven von M. domestica and O. aenescens. Angew. Parasitol. 23:143-154.

176

Müller, P., H. Schumann, P. Betke, H. H. Schultka, R. Ribbeck, and Th. Hiepe. 1981. Zur Bedeutung der *Musca domestica*-Antagonisten *Ophyra aenescens* (Diptera: Muscidae). I. Zum Auftreten von *Ophyra aenescens* in Anlagen der Tierproduktion. Angew. Parasitol. 22:212-216.

Pont, A. C. 1986. Family Muscidae, pp. 57-215. In: Soós, A., and L. Papp (eds.), Catalogue of Palaearctic Diptera, Vol 11. Akadèmiai Kiadò - Elsevier Sci. Publishers, Budapest, 346 p.

Schumann, H. 1982. Zur Bedeutung des *Musca domestica*-Antagonisten *Ophyra aenescens* (Diptera: Muscidae). II. Morphologie der Entwicklungsstadien. Angew. Parasitol. 24:113-114.

Sèguy, E. 1923. Diptères Anthomyides. In: Faune de France, 6: I-IX +393 p.

Skidmore, P. 1973. Notes on the biology of Palaearctic Muscids /1/, /2/. Entomologist 106:25-48, 49-59.

Stein, W., A. Gàl, and H. Gernath. 1977. Zum Auftreten von *Ophyra aenescens* (Wiedemann) (Dipt., Muscidae) in Deutschland, II., III. Z. Angew. Zool. 64:217-230, 311-324.

14. Coleopteran and Acarine Predators of House Fly Immatures in Poultry Production Systems

C. J. Geden

INTRODUCTION

Of all animal agricultural production systems, poultry best lends itself to the development of the biological control component of arthropod pest IPM programs (Axtell & Arends 1990). In particular, some caged layer systems present unique opportunities for filth fly biocontrol. Relatively stable environmental conditions within the houses, especially high relative humidity, promote the maintenance of pathogens of adult flies such as *Entomophthora muscae*. Similarly, the stability of the manure habitat and availability of hosts allows parasitoids of fly pupae to become established and assist in the regulation of pest fly species.

Although natural enemies have a considerable impact on the adult and pupal stages of flies, house fly eggs and young larvae are the life stages that are most vulnerable to biotic and abiotic factors that impinge on their development. Young house fly immatures are particularly susceptible to changes in manure quality such as moisture and must compete for resources with other pests such as *Ophyra* spp. and with non-pest species such as sphaerocerid flies, acarid mites, and nematodes. In warmer climates, larvae of the black soldier fly *Hermetia illucens* can modify the manure substrate so that it is no longer suitable for fly development, but this fly control is achieved at the cost of manure liquification and associated manure handling problems. When good management practices maintain manure in a relatively dry state (< 70% water) the long accumulation times typical of high-rise and deep-pit poultry houses often result in robust populations of coleopteran and acarine predators. Under proper conditions, these predators can maintain populations of pest fly species at nearly zero levels. As interest grows in the development of alternatives to chemical insecticides to manage fly populations, we can expect increased emphasis on the

177

development of fly predators into management tools. A thorough understanding of the ecology, behavior and life history of these organisms will be necessary if they are to be exploited successfully in IPM programs. Premature release attempts that do not take predator biology and the complexity of the manure ecosystem into account could adversely affect future acceptance of biological control by producers. The purpose of this chapter is to review the guild of beetle and mite predators in poultry manure and to identify areas where additional information is needed to support filth fly IPM programs.

PREDATOR SPECIES COMPOSITION AND SUCCESSION PATTERNS

Surveys in many regions of the world have demonstrated a wide variety of predatory and scavenger species of beetles and mites (Axtell 1963b, Anderson & Poorbaugh 1964, Peck and Anderson 1969, Axtell 1970a, Ito 1970, Legner & Olton 1970, Legner 1971, Legner et al 1973, 1975, Smith 1975, Toyama & Ikeda 1976, Pfeiffer & Axtell 1980, Geden & Stoffolano 1987, Corpuz-Raros & Sabio 1988, Hulley & Pfleiderer 1988). From these surveys, several broad conclusions may be drawn. First, the predator fauna in poultry manure is similar across broad geographic areas. Differences in species composition and richness generally are greater among production systems and over time within local areas than they are among different geographic regions. Second, in spite of the high species diversity of this guild, the great majority of individual predators belong to a small number of key species. Third, succession patterns in accumulating manure may follow predictable patterns. Fresh manure often is first colonized by parasitid mites (e.g., *Poecilochirus monospinosus* Wise, Hennessy & Axtell), followed by the principal species of histerid beetles (especially *Carcinops* spp.) and macrochelid mites (especially *Macrocheles muscaedomesticae* [Scopoli]), which persist in the accumulating droppings as long as prey are available. As the manure continues to age it is invaded by uropodid mites (e.g. *Fuscuropoda vegetans* [Foa]), staphylinid beetles, and the larger histerid and macrochelid species such as *Dendrophilus xavieri* Marseul and *Glyptholapsis confusa* (Foa). Among the predators that have been identified and observed, two species emerge as the most important predators of fly immatures. They are histerid beetles in the genus *Carcinops* and the macrochelid mite *Macrocheles muscaedomesticae. Carcinops pumilio* is typically found in temperate regions of the world, whereas *C. troglodytes* occurs in more tropical areas; the two species overlap in some localities (Wenzel 1955, Hulley & Pfleiderer 1988). Although *C. pumilio* and *M. muscaedomesticae* have very different life histories and behaviors, they both have high attack rates on fly immatures, both are present when fly pressure on the manure habitat is great, both occupy microhabitats that assure contact with house fly eggs, both have immature stages that feed on house flies, and both have mechanisms to leave the ecosystem when prey populations become sparse. Moreover, because both *M. muscaedomesticae* and *C. pumilio* feed on a variety of alternative prey as well as house fly immatures, high predator populations can be sustained even in the absence of the target pests. Because these two species are the best-studied and most important, their life histories will be discussed in separate sections later.

In addition to *M. muscaedomesticae*, two mite species that are often very abundant and can play important roles in the management of pest fly species are the parasitid *Poecilochirus monospinosus* and the uropodid *Fuscuropoda vegetans* (DeGeer).

P. monospinosus is a species that recently has been described from specimens collected from caged layer houses in North Carolina, where it is often present in high numbers in the first few weeks of manure accumulation (Wise et al. 1988). Adults and deutonymphs feed on house fly eggs and first instars. Female mites prefer first instars over fly eggs, and can destroy twenty-four fly immatures per day. Although the deutonymphs destroy fewer fly immatures on a daily basis (ca. 5) the deutonymphal stage is the longest in the ca. 17-day life cycle of the mite and is therefore the most abundant in field collections (Wise et al. 1988). Dispersal is accomplished phoretically by the deutonymph, and they will delay the final molt to the adult stage if fresh manure is not present. The mites feed readily on nematodes and acarid mites, and the volatile population dynamics of this species may be driven more by availability of these prey than by that of house fly immatures.

The geographic distribution of *P. monospinosus* remains uncertain because it was described only recently. The mites are often abundant in fresh manure in the spring and early summer in open-sided layer houses in North Carolina (Wise et al. 1988). Specimens have also been collected from enclosed houses in Massachusetts, and large populations have been observed in a deep pit house in February in New York (Geden et al. 1989) Additional investigations are necessary to define the geographic range of this species and to establish whether the mite occurs in other animal manures. Other parasitid mites occasionally are numerous in poultry and other manures, and careful examination of slide-mounted material is necessary for correct identification. Spot characters that can assist in the identification of *P. monspinosus* are a characteristically darkened band on the intercoxal shield (typical of most species of *Poecilochirus*) and the presence of a single membranous process on the fixed digit of the chelicerae.

Fuscuropoda vegetans occupies a niche that distinguishes it from other mites in the poultry manure ecosystem. The mites are slow moving and live well beneath the surface of the manure, where they feed on newly hatched fly larvae, nematodes, and other material in the manure (Willis & Axtell 1968). Their long development time (thirty to forty days), preference for fly larvae over eggs (O'Donnell & Axtell 1965) and preference for fairly dry manure (Peck and Anderson 1969) limits their utility in managing fly populations in the early weeks of manure accumulation. Very high population densities are observed in older manure deposits (Axtell 1970a, Legner et al. 1975), however, and in this more stable habitat the mites help destroy larvae that escape predation by other species that prefer the egg stage of the fly.

ALPHITOBIUS DIAPERINUS: A SPECIAL CASE

The lesser mealworm, *Alphitobius diaperinus* (Panzer) was known as a minor stored product pest and associate of the nests of bats and wild birds until the 1950's, when beetle populations were first noted in large numbers in poultry houses (Gould & Moses 1951, Harding & Bissell 1958). Now cosmopolitan, the original range of this species was tropical Africa. The advent of high-density poultry production in the years after World War II provided this tenebrionid with an ideal habitat in the form of large volumes of avian droppings plus the opportunity to feed on spilled poultry feed.

The beetles feed on manure, spilled feed, other insects, and cadavers in the litter of most types of poultry housing. When populations become high, mature beetle larvae tunnel into thermal insulation materials in the houses in search of protected pupation sites (Ichinose et al. 1980, Le Torc'h & Letenneur 1983, Vaughan et al. 1984, Geden & Axtell 1987), resulting in as much as $20,000 worth of damage to individual houses (Turner 1986). Adult beetles cause public nuisance problems by invading neighboring dwellings when infested litter is applied to fields after house cleanouts

(Thornberry 1978, Smith 1981). The adults will also emigrate from poultry houses at night and are attracted to light, including the electrocuting black light devices that people use in attempts to manage biting fly problems near their homes. The species of *Alphitobius* are somewhat difficult to distinguish (Green 1980), and it is common to assume that *A. diaperinus* accounts for most infestations because of historical precedent. Other species of *Alphitobius*, such as the black fungus beetle *A. laevigatus* (Fab.) also occur in poultry production systems (cited in Preiss & Davidson 1970), and more species may be introduced in the years ahead. Careful examination of representative beetles from infestations occasionally should be made so that new species introductions can be detected before they become widespread.

In spite of their destructive and nuisance-causing behaviors, the beetles perform beneficial functions in the manure of caged layer houses. The activity of the beetles tunnelling through the manure assists greatly in aeration and drying, thus reducing opportunities for fly breeding. Adult and larval beetles are also facultative predators and will attack house fly eggs and larvae in the laboratory (Toyama & Ikeda 1976, Despins et al. 1988). Because of these beneficial activities it has been suggested that poultry producers encourage high populations of lesser mealworm for manure and fly management (Wallace et al. 1985). Successful use of *A. diaperinus* in this fashion would require the development of means to limit their destructive activities. Premise treatments with insecticides to protect insulation are effective for only a short time because of the rapid accumulation of dust on the surfaces. Until effective barriers to beetle movement out of the pit are developed, this beetle will remain a severe pest of the poultry industry.

LIFE HISTORY OF *CARCINOPS PUMILIO*

Carcinops pumilio is the most common histerid found in poultry manure throughout much of the world. This species probably evolved as an associate of the nests of wild birds in Africa. An opportunistic predator, *C. pumilio* has been collected from many habitats other than avian manure including cut grass, stored grain, stale yeast, glue factories, & carrion (Hinton 1945). I have also collected them from calf bedding on New York dairy farms and from an experimental composting toilet. *C. pumilio* prefers dark, protected habitats and does not occur in dung pats on pasture. A second *Carcinops* species, *C. troglodytes*, also occurs in warmer regions of the world, and the two species may overlap in some areas. Because the two are very similar in appearance, some locality records for both species are suspect; additional surveys are needed to define their distribution patterns more clearly (Wenzel 1955). Another histerid that can be confused with *C. pumilio* is *Gnathoncus nanus*, which is sometimes quite common in poultry houses in the United States (Pfeiffer & Axtell 1980). *C. pumilio* is the only *Carcinops* species recorded from poultry manure in temperate North America, and our knowledge of the life history of this important genus essentially is limited to a few North American studies of *C. pumilio*.

Feeding Habits of Adults and Larvae

Adult beetles and both larval instars are predaceous. In poultry houses, adults feed preferentially on house fly and other large muscoid fly eggs and newly hatched larvae. Individual beetles are unable to destroy second and third instar house fly larvae, although groups of beetles will attack and eventually destroy these larger prey. They will also attack newly emerging flies during eclosion and tear at the ovipositor of adult flies when they attempt to oviposit in beetle-rich environments. In addition to house fly prey, adults feed readily on acarid mites and other, small dipterans, especially larvae of small dung flies (family Sphaeroceridae). It is not known whether they will feed on saprophytic nematodes, but female beetles that are held with nematodes as their only prey source will not develop eggs.

The feeding habits of the larvae are not well understood. Both larval instars feed on house fly eggs, although the first instars are so small that their contribution to fly control is probably minor. Both larval stages also feed readily on acarid mites, saprophytic nematodes, and sphaerocerid immatures. No data are available on the relative nutritional quality of these prey for the developing beetle immatures, nor is it known whether the larvae show any preferences for one type of prey over another.

Fecundity, Development and Longevity

The following life history description is summarized from data in Smith (1975), Morgan et al. (1983), Geden (1984), and unpublished personal observations. Individually held females produce an average of 1.8 eggs per day when they are provided with frozen house fly eggs as prey and sand as an oviposition substrate. When living immature sphaerocerid prey are used in a seminatural substrate, oviposition rates can exceed ten eggs per beetle per day. All mobile life stages of *C. pumilio* are highly cannibalistic, and apparent fecundity (net production of living larvae) drops off sharply with modest increases in beetle densities, even when prey are abundant. Cannibalism effects are detectable at crowding levels of as few as 10 beetles per 1000 cm^3 of medium. Under field conditions, where beetle densities in excess of 100 adults per 1000 cm^3 are common, cannibalism is likely one of the principal factors that limit the upper bounds of beetle populations.

As indicated above, the adults of *C. pumilio* are opportunistic predators that feed readily on acarid mites and other dipterans. They also feed on dead arthropods, bird carcasses, and cracked chicken eggs. No data are available on the relative contributions of different types and quantities of food to the fecundity of this species.

Eggs are deposited in cracks and crevices in the manure, and are extremely difficult for human investigators to locate. Feeding larvae are typically found in moist clumps of manure just under the manure surface, where newly hatched house fly larvae and sphaerocerid prey are most abundant. Mature second instars, after purging the alimentary tract, form a pupation cell from bits of debris in the manure. They will also pupate in soil and in empty fly puparia if these sites are available. Newly eclosed beetles spend the first few days of adult life in the pupal cell then emerge after the cuticle is fully hardened. Development from egg to adult is completed in twenty-one to forty days at 25-30°C. The second instar is the longest-lived individual life stage, and accounts for nearly one half of the total development time. Two-thirds of the duration of the second instar is spent in a post-feeding prepupal stage. The pupal stage accounts for about one third of the development time and is the stage that suffers the least mortality.

The sex ratio at emergence is close to 1:1. Females are responsive to mating immediately after eclosion, whereas males are not capable of mating until they have fed on prey for about one week. A spermatophore is transferred to the common oviduct during copulation, an act that is rarely observed. Little is known of the mating behavior, nor is it known how frequently females must be reinseminated to maintain fertility. The beetles will feed on prey by two days post-emergence, and will survive for about eleven days after emergence if no prey are available. In contrast, older beetles with a prey-rich feeding history can survive for ca. 40 days without consuming prey.

There are four polytrophic ovarioles per ovary. Yolk deposition in at least one ovariole is visible by three days post-emergence, and the first egg is deposited one to two weeks after emergence. Eggs are developed singly such that one rarely finds more than two follicles per ovary with substantial yolk deposition. Single, mature eggs fill much of the body cavity. Mating is not necessary for egg maturation, but virgin females rarely oviposit.

On average, both male and female beetles live for about three months (at 30°C) if prey are abundant. Survival rates and patterns for the sexes are very similar in the laboratory, but field populations are significantly biased toward males. The beetles can withstand prolonged periods of starvation between bouts of prey-feeding, and will live for over a year if prey are alternately available and unavailable. Temperature effects on adult survival are not known, but beetles can be stored in an inactive state at 10°C for several months with little mortality if proper moisture conditions are maintained. There is no evidence for diapause in any stage in commercial poultry production facilities. Beetle populations persist in the manure of high-rise poultry houses throughout the winter months, and all beetle stages can be collected at any time of the year. Investigations of beetles under feral conditions may reveal overwintering mechanisms that are not manifest in modern poultry houses.

Dispersal

C. pumilio adults typically do not fly, are repelled by strong light, and live just under the surface of the manure. Occasionally, large numbers of beetles can be seen climbing the walls towards lights and exhaust fans and flying out of production facilities. These beetles are no different from those under the manure surface with respect to body size, sex ratios, mating status, or physiological age. The phototactic and flight initiation behaviors of such field-collected "dispersers" can be reversed by the administration of dipteran prey in the laboratory. Conversely, beetles with a prey-rich feeding history can be induced to fly by depriving them of prey for four to five days (Geden et al. 1987).

Given the temporary nature of the wild habitats that the beetles inhabited before the domestication of birds, selection for a means of assessing and fleeing deteriorating prey conditions may be assumed to have been strong. In modern poultry production systems these same behaviors provide a "safety valve" for growing beetle populations and assist the establishment of new predator populations in houses that have been cleaned out.

LIFE HISTORY OF *MACROCHELES MUSCAEDOMESTICAE*

In contrast with *C. pumilio*, the literature on *M. muscaedomesticae* is voluminous, and this mite is one of the best-studied natural enemies of filth flies. It is cosmopolitan in its distribution and is found in the manure of swine, cattle, sheep, and ducks as well as that of chickens (Axtell 1963b, Ito 1970). Although there are other macrochelid species that are common in poultry manure, especially *Glyptholapsis confusa* and *M. merdarius* (Berlese), *M. muscaedomesticae* is the most abundant macrochelid in caged layer houses when fly activity in the manure is greatest. The importance of this mite in regulating populations of filth flies has long been recognized (Pereira & de Castro 1945), and this topic has been reviewed by Anderson (1983), Axtell (1969, 1970a, 1981, 1986), and Krantz (1983).

Feeding Habits of Adults and Nymphs

Adult female mites feed preferentially on house fly eggs over first instar larvae (O'Donnell & Axtell 1965, Rodriguez & Wade 1961); adult males have received little attention but have a negligible impact on developing fly immatures (personal observations). Females also feed on immatures of other fly pests such as the stable fly, lesser house fly, face fly, and blow flies, and on sphaerocerid immatures, especially *Coproica hirtula* (Rondani). Other arthropod prey include acarids and other mites.

Saprophytic nematodes are a critical food item that can regulate mite population dynamics in the field. *M. muscaedomesticae* can be successfully raised using nematodes as the sole food source (Filipponi & Delupis 1963) and mass-rearing methods for this mite that employ nematodes hold substantial promise for future release experiments (Ho et al., elsewhere in this volume). Adult mites prefer fly eggs over nematodes but the nymphs prefer nematodes over fly eggs (Rodriguez et al. 1962, Ito 1977b). Prey preferences are graded responses that are tempered by the relative availability and abundance of different prey, and results of preference experiments should be viewed with caution. For example, even though the adult mites may have an innate preference for fly eggs over nematodes, the presence of high nematode populations will deflect substantial mite feeding away from fly eggs (Geden & Axtell 1988).

Fecundity, Development and Longevity

Females lay one to twenty-five eggs per day; mite density, type of prey, mating status, substrate, temperature, and physiological age of the mites all influence fecundity. Nematodes may be equal, superior, or inferior to fly eggs for mite reproduction, depending on the species of nematode under consideration (Rodriguez et al. 1962, Ito 1977b). The mites are arrhenotokous, resulting in a sex ratio that is somewhat male-biased (ca. 40-45% females [Cicolani 1979, Wade & Rodriguez 1961]). Because female mites live longer than males, the sex ratio at oviposition becomes masked over time so that field collections or older laboratory cultures give the impression of female rather than male bias (Ito 1977a, Stafford &Bay 1987).

M. muscaedomesticae is highly cannibalistic (Filipponi & Petrelli 1967). Like *C. pumilio*, cannibalism is evident at relatively low mite densities. Apparent fecundity (production of new adults) declines by 40% when mite densities (with an overabundance of prey) increase from 6 to 20 females per 1000 cm^3 of medium (Geden et al. 1990). Since mite densities in the field commonly exceed 50 mites per 1000 cm^3 of manure, cannibalism plays an important role in limiting mite population sizes. This factor may account for some of the variation in fecundity reported in the literature and for the boom/bust population dynamics of the this predator in the field (Axtell 1970a,b, Geden & Stoffolano 1987).

The entire life cycle (egg to adult) of *M. muscaedomesticae* is completed in forty-two to fifty-four hours at temperatures that are typical of summer conditions in poultry manure (27-30°C [Wade & Rodriguez 1961, Cicolani 1979]). Newly emerged females begin ovipositing about two days after the final moult at 27°C, continue to oviposit for twelve days, then may survive as post-reproductive individuals for an additional ten days (total longevity = ca. 24 days). Males live for about fifteen days. Filipponi & Petrelli (1967) provided detailed data on the influence of temperature and mite age on fecundity, development and longevity of *M. muscaedomesticae*;

results of these studies are presented in summarized form in a computer simulation model of the mite (Geden et al. 1990).

Dispersal

M. muscaedomesticae disperses from deteriorating habits by phoretic transport of adult mites on flies (see reviews by Axtell [1969] and Farish & Axtell [1971]; Jalil & Rodriguez 1970, Borden 1989). House flies are the most common phoretic hosts, although the mites will attach to other muscoid flies such as *Fannia* and *Ophyra* spp. As the manure dries and becomes less suitable for fly oviposition and development (Sukarsih et al. 1989), females become increasingly attracted to flies that continue to visit the manure. Sensillae on the first pair of tarsi assist the mite in orienting towards olfactory cues given off by living flies; dead flies remain attractive for a short time after death. The mite then makes an assessment of the condition of the host by sensillae on the palps and attaches to it by the chelicerae. The mites generally do not have an adverse effect on the flies, although some flies may become so burdened with these and other phoretic mites that normal walking and flight behaviors are impeded (Elzinga & Broce 1988) . The mites remain attached to the host until the fly visits fresh manure that is suitable for fly and mite development. The chemical cues involved in mite attachment and detachment have not been identified. Ammonia has been suggested as a detachment stimulus in some studies (Pareira & de Castro 1947, Jail & Rodriguez 1970) but not in others (Farish & Axtell 1971). Dry manure that previously stimulated attachment to flies can stimulate detachment if it is remoistened to levels close to those of fresh droppings (Farish & Axtell 1971).

ECOLOGICAL CONSIDERATIONS

C. pumilio and *M. muscaedomesticae* are sensitive to changes in manure quality, especially moisture. *C. pumilio* adults are present in manure with anywhere from 10 to 70% moisture, whereas the larvae are most abundant in manure in the 50-70% moisture range (Geden & Stoffolano 1988). In general, high beetle populations are favored by relatively dry manure (Peck & Anderson 1969, Bills 1973, Smith 1975), and predator performance is poor in manure with greater that 70% moisture. Adult *M. muscaedomesticae* occupies a somewhat narrower band of manure tolerance and is most common in manure with 50-70% moisture (Geden & Stoffolano 1988, Stafford & Bay 1987). Data are not available on attack rates at different manure moisture levels, but there appears to be a window of vulnerability in the 70-80% moisture range where the manure is suitable for fly oviposition and development but is too moist for the predators to forage effectively.

Manure age also influences predator activity, and it has long been suggested that old manure be left behind at cleanout to allow predators to recolonize fresh droppings. This approach is effective for the mites, which rapidly colonize prey-rich fresh droppings from old manure. The beetles, however, show a strong preference for older manure over fresh droppings, even when prey are much more abundant in the latter (Geden & Stoffolano 1988). Very fresh manure ($\leq$2 weeks accumulation) appears to have a repellent effect on *C. pumilio*, although this may be due to the high water content of new droppings (ca. 75%) rather than any specific chemical repellency. It remains to be determined whether such fresh manure can be manipulated in a manner that would make it more acceptable to the beetles in predator release or conservation programs.

Large predator populations of both species often are present in large manure deposits of long accumulation times. The beetles and mites in such older accumulations only inhabit a relatively narrow band of the manure near the surface (Willis & Axtell 1968, Geden & Stoffolano 1988, Stafford et al. 1988). Even though *C. pumilio* avoids freshly accumulating manure in recently cleaned out houses, both the beetles and mites are most abundant near the crest of the manure row, where fresh droppings attract ovipositing flies. Attempts to conserve predators during cleanouts may be made more effective by selectively harvesting the predator-rich manure surface and scattering this material under cleaned-out cage rows.

The poultry manure community is a complex system whose population dynamics and interspecific interrelationships are poorly understood. Predator populations over time show a degree of volatility that can not be explained solely on the basis of availability of known prey items in the habitat (Peck and Anderson 1969, Axtell 1970a, Legner et al. 1975, Geden & Stoffolano 1987). *M. muscaedomesticae* populations typically increase rapidly, reach very high population levels, then crash to lower, more stable densities. Similar boom/bust patterns occur with *C. pumilio* larvae, whereas adult beetle densities increase slowly to high, more stable levels. Both species are opportunistic predators and cannibals that feed on a variety of alternate prey when pest fly species are absent. *M. muscaedomesticae* also attacks *C. pumilio* immatures in the laboratory and greatly suppresses beetle populations in culture. Interspecific predation as well as competition for prey undoubtedly limits populations of both predators in the field.

In addition to direct interspecific interactions, other manure inhabitants can alter the physical characteristics of the manure in such as way as to depress predator populations. In warmer climates, larvae of *H. illucens* modify the consistency of the manure so that it takes on a fluid consistency. Intense house fly larval activity breaks up the discreet clumps of manure that are needed for *C. pumilio* oviposition and pupal survival. High population densities of lesser mealworms also disrupt clumps of manure, and their activity creates conditions that are inimical to the maintenance of alternate prey of the predators, especially nematodes and sphaerocerids. Much additional research is needed on these interactions to define in a quantitative manner the ecological circumstances that most favor predator persistence and effectiveness.

ATTACK RATES

Beetle and mite attack rates on house fly immatures vary depending on methods of testing. The range in attack rates that have been observed in laboratory assays are presented in Table 1, along with those of other species for purposes of comparison (reviews on this subject are in Axtell 1969, Geden et al. 1988, and Geden & Axtell 1988). *C. pumilio* adults have been reported to destroy between thirteen and over 100 fly immatures per beetle per day. These extremes represent tests using sated beetles offered frozen eggs and starved beetles offered fresh eggs, respectively. Using standardized testing methods with natural substrates and live house fly prey, the following general observations may be made about *C. pumilio* attack rates: 1) beetles in the normal hunger range at 27°C kill forty to fifty-five fly immatures per day if they are provided with an abundance of fly eggs; 2) sated beetles kill twenty-five fly immatures per day, whereas starved beetles kill over 100; 3) attack rates are unaffected by beetle crowding at densities typical of field populations; 4) attack rates range from twelve to eighty-three fly immatures destroyed per day over a range of temperatures from 15 to 33°C; 5) attack rates are depressed in the presence of acarid mites but not nematodes or sphaerocerid flies; 6) male and female beetles have similar attack rates; 7) attack rates are fairly constant over different experimental substrates; and, 8) second instar larvae kill ca. 26 fly immatures per day.

Table 1. Attack rates (no. house fly immatures destroyed per predator per day) of common beetle and mite predators in laboratory assays.

Carcinops pumilio adults	13-103
Carcinops pumilio second instars	13-26
Dendrophilus xavieri adults	4-13
M. muscaedomesticae females	3-25
Glyptholapsis confusa females	5-10
Poecilochirus monospinosus females	2-10
Poecilochirus monospinosus nymphs	2-5
Fuscuropoda vegetans adults	1-3

Adult *M. muscaedomesticae* attack rates in the literature also vary considerably. The following generalizations may be made: 1) mites under most conditions at 27°C destroy ten to twenty fly immatures per day; 2) hunger levels have little impact on daily rates of house fly destruction; 3) mites are affected greatly by crowding, and small increases in mite densities result in large decreases in attack rates; 4) attack rates range from five to thirty-six fly immatures destroyed per day over a range of temperatures between 15° and 33°C; 5) attack rates are suppressed in the presence of nematodes and sphaerocerids but not acarids; 6) attack rates by males are negligible; and, 7) attack rates are influenced greatly by the test substrate and associated olfactory stimuli.

Addition of a second predator species to a one-predator arena indicates that the attacks of two species are not always additive. *C. pumilio* and *M. muscaedomesticae* attacks are additive as long as the mites are not given greatly superior numerical advantage. When they are (eg., thirty mites to one beetle), there can be a significant depressing effect on attacks owing to mite interference with the foraging behavior of the beetles. Attack rates of *M. muscaedomesticae* and the parasitid *Poecilochirus monospinosus* are additive except when the parasitids are given substantial numerical advantage. In contrast, the parasitid mites interfere with *C. pumilio* attacks in every combination that has been tested (Geden et al. 1988). Thus it is difficult to estimate what the actual attack rates of these organisms may be in the field, where many levels of interspecific interactions can affect their behavior and efficacy as biological control agents.

Relatively little information is available on predator effectiveness in the field. Several studies have demonstrated up to 99% reductions in fly emergence from manure containing either naturally occurring or released predators (Axtell, 1963a, Singh et al. 1966, Rodriguez et al. 1970, Legner 1971, Propp & Morgan 1985, Geden et al. 1988). These investigations indicate that attack rates under field conditions are somewhat lower than are observed in the laboratory. Field-derived attack rates for *M. muscaedomesticae* range from 2.3 (Singh et al. 1966) to 7.5 (Geden et al. 1988) fly immatures destroyed per day, whereas *C. pumilio* adults and larvae destroy thirty-seven and seventeen and fly immatures per day, respectively. Additional investigations are needed to evaluate the effects of manure moisture, age and consistency on predator effectiveness in the field.

PROSPECTS FOR USE IN MANAGEMENT PROGRAMS

C. pumilio and *M. muscaedomesticae* compliment one another in regulating populations of house flies and other pest species (Table 2). The mites invade fresh manure when fly oviposition pressure is greatest in the habitat. Mite populations are volatile and increase rapidly in the early weeks of manure accumulation. In contrast, the beetles are slow to colonize fresh manure accumulations but build slowly to high population densities if proper manure conditions are maintained. The mites kill fewer fly immatures per capita than do the beetles, but they are often more numerous. The two predators prefer somewhat different microhabitats, and both show broad ecological overlap with pest species. Other predators that have received less attention here and elsewhere also contribute to total fly mortality. Parasitid mites attack fly immatures in very fresh manure where *C. pumilio* is lacking, and uropodids deeper in the manure attack fly larvae that escape predation by surface-dwelling predators that prefer the eggs. Minor coleopteran predator species such as *D. xavieri* and certain staphylinids can play important roles in managing flies, especially in older manure that is less attractive to *M. muscaedomesticae*.

The manure arthropod community is complex and efforts to manipulate predators in IPM programs must take this complexity into account. Predator releases may have potential as a management

Table 2. Comparison of *Carcinops pumilio* and *Macrocheles muscaedomesticae*:

Attribute	*C. pumilio*	*M. muscaedomesticae*
Development time	21-40 days	2-3 days
Attack rate (no. flies killed/day)[1]	35-40	5-10
Adult longevity	90 days-2 years	10-25 days
Max. fecundity (eggs/day)	10	10
Relative abundance	less numerous	more numerous
Volatility of populations	relatively stable	highly volatile
Manure age preference	older manure	fresh manure
Manure moisture preference	10-70% water	50-70% water
Main alternate prey	sphaerocerids, acarids	nematodes, sphaerocerids
Prey of immatures	fly, acarine, nematode	nematode, fly, acarine
Cannibalism	yes	yes
Parasites/pathogens	none known	none known
Dispersal	hunger>flight	manure drying> phoresy

[1] Estimates from field experiments.

tool, but only under certain circumstances. Mass-releases of single species into older, established manure accumulations would likely have little impact on fly control. Releases can only be expected to be effective as part of a rigorous overall management program. Leaking waterers, poor drainage and inadequate ventilation result in manure that is too wet for these beneficials. Similarly, predators should not be released into manure that is already overrun with fly larvae. Rather, release efforts should be aimed at repopulating houses after cleanout in cases where sanitation concerns prohibit leaving residues of old manure. Releases in such cases should be conducted in two phases, with a release of *M. muscaedomesticae* within a few days of cleanout followed by a *C. pumilio* release three weeks later. No data are available on appropriate numbers for such inoculative releases, but twenty mites and ten beetles per cage should be sufficient.

Several obstacles must be overcome before releases will be practical as a routine matter. Predators can be harvested from poultry houses where they abundant, but inter-farm movement of these organisms poses biosecurity problems. Economical mass-rearing methods need to be developed. The house fly is an unsuitable prey source for mass-production because larvae that escape predation disrupt the substrate. Alternate prey can be used, such as nematodes for the mites (Ho et al., elsewhere in this volume) and sphaerocerid flies for the beetles (Geden 1984), but current rearing methods remain relatively costly. Cost-effective predator production must await the development of artificial diets and methods of handling that will minimize the effects of cannibalism in culture.

At present, predator conservation is more practical than augmentation. Naturally occurring beetle and mite populations can be conserved and promoted by good management practices. Manure should be kept dry by maintaining good ventilation, repairing drainage problems, and attending to leaking waterers promptly. Larviciding should be avoided unless the insecticide is known to have no adverse effect on beneficials (Axtell 1970b, Axtell & Edwards 1983, Meyer et al. 1984). A residue of old manure should be left behind after cleanout to allow predator colonization of fresh droppings. Scattering old manure under the cage rows at cleanout might accelerate predator establishment. Fly population growth in the weeks after cleanout should be managed using methods of insecticide application that will do minimal harm to the predators. This can be accomplished by the use of insecticides that have demonstrably low toxicities to beetles and mites such as cyromazine or by restricting broad spectrum insecticide applications to fly resting areas away from the manure surface. Populations of sphaerocerid flies, although annoying at times, are important promoters of predator populations and should not be controlled.

In summary, the guild of coleopteran and acarine predators is a potent force in the natural regulation of filth fly populations in poultry houses. Their full incorporation into IPM programs as a management tool requires a better understanding of the complex interactions of the manure ecosystem. Future research should address the problems of quantifying the ecological parameters that are necessary for predator establishment and efficacy. Simulation models can help us to understand some of these factors and to identify areas where greater empirical information is needed. Additional research is also needed to determine whether and in what

194

combinations predator releases are effective, and to develop cost-effective rearing methods for these beneficials.

REFERENCES CITED

Anderson, J. R. 1983. Mites as biological control agents of dung-breeding pests: practical considerations and selection for pesticide resistance. Pages 99-102 in M. A. Hoy, G. L. Cunningham, and L. Knutson (eds). 1983. Biological Control of Pests by Mites. University of California Agricultural Experiment Station Special Publication 3304.

Anderson, J. R. and J. H. Poorbaugh. 1964. Observations on the ethology and ecology of various Diptera associated with northern California poultry ranches. J. Med. Entomol. 1: 131-147.

Axtell, R. C. 1961. New records of North American Macrochelidae (Acarina: Mesostigmata) and their predation rates on the house fly. Annals Entomol. Soc. Amer. 54: 748.

_____. 1963a. Effect of Macrochelidae (Acarina: Mesostigmata) on house fly production from dairy cattle manure. J. Econ. Entomol. 56: 317-321.

_____. 1963b. Acarina occurring in domestic animal manure. Annals Entomol. Soc. Amer. 56: 628-633.

_____. 1969. Macrochelidae (Acarina: Mesostigmata) as biological control agents for synanthropic flies. Pages 401-406 in Proc. 2nd Int. Congr. Acarol. 1967. Akademai Kiado, Budapest.

_____. 1970a. Integrated fly control program for caged-poultry houses. J. Econ. Entomol. 63: 400-405.

_____. 1970b. Fly control in caged-poultry houses: Comparison of larviciding and integrated control programs. J. Econ. Entomol. 63: 1734-1737.

_____. 1981. Use of parasites and predators in filth fly IPM programs. In: Proceedings of a Workshop on Status of Biological Control of Filth Breeding Flies. U.S. Department of Agriculture A106.2:F64.

_____. 1986. Fly management in poultry production: cultural, biological, and chemical. Poultry Sci. 65: 657-667.

Axtell, R. C. and J. J. Arends. 1990. Ecology and management of arthropod pests of poultry. Ann. Rev. Entomol. 35:101-126.

Axtell, R. C., and T. D. Edwards. 1983. Efficacy and nontarget effects of Larvadex as a feed additive for controlling house flies in caged-layer poultry manure. Poultry Sci. 62: 2371-2377.

Bills, G. T. 1973. Biological fly control in deep-pit poultry houses. British Poultry Sci. 14: 209-212.

Borden, E. E. R. 1989. The phoretic behavior and olfactory preference of *Macrocheles muscaedomesticae* (Scopoli) (Acarina: Macrochelidae) in its relationship with *Fannia canicularis* (L.) (Diptera: Muscidae). Pan-Pacific Entomol. 65: 89-96.

Cicolani, B. 1979. The intrinsic rate of natural increase in dung macrochelid mites, predators of *Musca domestica* eggs. Boll. Zool. 46: 171-178.

Corpuz-Raros, L. A., G. C. Sabio and M. Velasco-Soriano. 1988. Mites associated with stored products, poultry houses and house dust in the Philippines. Philipp. Entomol. 7: 311-321.

Despins, J. L., J. A. Vaughan and E. C. Turner, Jr. 1988. Role of the lesser mealworm, *Alphitobius diaperinus* (Panzer) (Coleoptera: Tenebrionidae) as a predator of the house fly, *Musca domestica* L. (Diptera: Muscidae) in poultry houses. Coleopt. Bull. 42: 211-216.

Elzinga, R. J. and A. B. Broce. 1988. Hypopi (Acari: Histiostomatidae) on house flies (Diptera: Muscidae): a case of detrimental phoresy. J. Kansas Entomol. Soc. 61: 208-213.

Farish, D. J. and R. C. Axtell. 1971. Phoresy redefined and examined in *Macrocheles muscaedomesticae* (Acarina: Macrochelidae). Acarologia 13: 16-29.

Filipponi, A. and D. di Delupis. 1963. Sul regime dietetico di alcuni macrocheliidi (Acari: Mesostigmata) associata in natura a Muscidi di interesse sanitario. Riv. Parassitol. 24: 277-288.

Filipponi, A. and M. G. Petrelli. 1967. Autecologia capacita' moltiplicativa di *Macrocheles muscaedomesticae* (Scopoli) (Acari: Mesostigmata). Riv. Parassitol. 28:129-156.

Geden, C. J. 1984. Population dynamics, spatial distribution, dispersal behavior and life history of the predaceous histerid, *Carcinops pumilio* (Erichson), with observations of other members of the poultry manure arthropod community. Ph.D. dissertation, Univerity of Massachusetts, Amherst.

Geden, C. J. and R. C. Axtell. 1987. Factors affecting climbing and tunneling behavior of the lesser mealworm, *Alphitobius diaperinus* (Coleoptera: Tenebrionidae). J. Econ. Entomol. 80: 1197-1204.

Geden, C. J. and R. C. Axtell. 1988. Predation by *Carcinops pumilio* (Coleoptera: Histeridae) and *Macrocheles muscaedomesticae* (Acarina: Macrochelidae) on the house fly (Diptera: Muscidae): functional response, and effects of temperature and availability of alternative prey. Environ. Entomol. 739-744.

Geden, C. J. and J. G. Stoffolano, Jr. 1987. Succession of manure arthropods at a poultry farm in Massachusetts, with notes on *Carcinops pumilio* sex ratios, ovarian condition and body size. J. Med. Entomol. 24: 214-222.

_____. 1988. Dispersion patterns of arthropods associated with poultry manure in enclosed houses in Massachusetts: spatial distribution and effects of manure moisture and accumulation time. J. Entomol. Sci. 23: 136-148.

Geden, C. J. , J. G. Stoffolano, Jr. and J. S. Elkinton. 1987. Prey-mediated dispersal behavior of the predaceous histerid, *Carcinops pumilio*. Environ. Entomol. 16: 415-419.

Geden, C. J., R. E. Stinner and R. C. Axtell. 1988. Predation by predators of the house fly in poultry manure: effects of predator density, feeding history, interspecific interference, and field conditions. Environ. Entomol. 17: 320-329.

Geden, C. J., D. C. Steinkraus, and D. A. Rutz. 1989. *Poecilochirus monospinosus* (Acarina: Parasitidae), a predator of house fly immatures: new locality records from New York and Massachusetts. J. New York Entomol. Soc. (in press).

Geden, C. J., R. E. Stinner, D. Kramer and R. C. Axtell. 1990. MACMOD, a FORTRAN simulation model of *Macrocheles muscaedomesticae* population dynamics and predation on house fly immatures. Environ. Entomol. 18: (in press).

Gould, G. E. and H. E. Moses. 1951. Lesser mealworm infestation in a brooder house. J. Econ. Entomol. 44: 265.

Green, M. 1980. *Alphitobius viator* Mulsant & Godart in stored products and its identification (Coleoptera: Tenebrionidae). J. Stored Prod. Res. 16: 67-70.

Hinton, H. E. 1945. The histeridae associated with stored products. Bull. Entomol. Res. 35: 309-340.

Hulley, P. E. and M. Pfleiderer. 1988. The Coleoptera in poultry manure - potential predators of house flies, *Musca domestica* Linnaeus (Diptera: Muscidae). J. Entomol. Soc. South Africa 51: 17-29.

Ichinose, T., S. Shibazaki and M. Ohta. 1980. Studies on the biology and mode of infestation of the tenebrionid beetle *Alphitobius diaperinus* (Panzer) harmful to broiler-chicken houses. Jpn. J. Applied Zool. 34: 417-421.

Ito, Y. 1970. Preliminary surveys on macrochelid and some other mesostigmatid mites occurring in the experimentally deposited live-stock dungs as predators of muscid flies. Jpn. J. Sanit. Zool.21: 205-208.

_____. 1977a. Changes of the population density and stage compositions of three mesostigmatid mite species on a restricted food supply. Jpn. J. Appl. Enomol. Zool. 21: 74-78.

_____. 1977b. Predatory activity of mesostigmatid mites (Acarina: Mesostigmata) for house fly eggs and larvae under feeding of nematodes. Jpn. J. Sanit. Zool. 28: 167-173.

Jalil, M. and J. G. Rodriguez. 1970. Studies of behavior of *Macrocheles muscaedomesticae* (Acarina: Macrochelidae) with emphasis on its attraction to the house fly. Annals Entomol. Soc. Amer. 63: 738-744.

Krantz, G. W. 1983. Mites as biological control agents of dung-breeding flies, with special reference to the Macrochelidae. Pages 91-98 in M. A. Hoy, G. L. Cunningham, and L. Knutson (eds). 1983. Biological Control of Pests by Mites. University of California Agricultural Experiment Station Special Publication 3304.

Legner, E. F. 1971. Some effects of the abmient arthropod complex on the density and potential parasitization of muscoid Diptera in poultry wastes. J. Econ. Entomol.64: 111-115.

Legner, E. F. and G. S. Olton. 1970. Worldwide survey and comparison of adult predator and scavenger insect populations associated with domestic animal manure where livestock is artificially congregated. Hilgardia 40: 225-266.

Legner, E. F., W. R. Bowen, W. D. McKeen, W. F. Rooney and R. F. Hobza. 1973. Inverse relationship between mass of breeding habitat and synanthropic fly emergence and the measurement of population densities with sticky tapes in California inland valleys. Environ. Entomol. 2: 199-205.

Legner, E. F., G. S. Olton, R. E. Eastwood and E. J. Dietrick. 1975. Seasonal density, distribution and interactions of predatory and scavenger arthropods in accumulating poultry wastes in coastal and interior southern California. Entomophaga 20: 269-283.

Le Torc'h, J. M. and R. Letenneur. 1983. Laboratory tests of resistance of different thermic insulators to the boring of the tenebrionid *Alphitobius diaperinus* (Col. Tenbrionidae). Comptes Rendus Hebdomadairs des Sceances, Acad. Agric., Paris 806: 188-200.

Meyer, J. A., W. F. Rooney and B. A. Mullens. 1984. Effect of Larvadex feed-through on cool-season development of flith flies and beneficial Coleoptera in poultry manure in southern California. Southwestern Entomol. 9: 52-55.

Morgan, P. B., R. S. Patterson and D. E. Weidhaas. 1983. A life-history study of *Carcinops pumilio* (Erichson) (Coleoptera: Histeridae). J. Georgia Entomol. Soc. 18: 353-359.

O'Donnell, A. E. and R. C. Axtell. 1965. Predation by *Fuscuropoda vegetans* (Acarina: Uropodidae) on the house fly (*Musca domestica*). Annals Entomol. Soc. Amer. 58: 403-404.

Peck, J. H. 1969. Arthropod predators of immature Diptera developing in poultry droppings in northern California. Part II. Laboratory studies on feeding behavior and predation potential of selected species. J. Med. Entomol. 6: 168-171.

Peck, J. H. and J. R. Anderson. 1969. Arthropod predators of immature Diptera developing in poultry droppings in northern California. Part I. Determination of seasonal abundance and natural cohabitation with prey. J. Med. Entomol. 6: 163-167.

Pereira, C. and M. P. de Castro. 1945. Contribucao para o conhecmento da especie tipo de "*Macrocheles* Latr." ("Acarina"): "*M. muscaedomesticae* (Scopoli, 1772)". Arch. Inst. Bio., 16: 153-186.

_____. 1947. Forese e partenogese arrenotoca em "*Macrocheles muscaedomesticae*" (Scopoli) ("Acarina": Macrochelidae) e sua signifcacao ecologica. Arch. Inst. Bio., 18: 71-89.

Pfeiffer, D. G. and R. C. Axtell. 1980. Coleoptera of poultry manure in caged-layer houses in North Carolina. Environ. Entomol. 9: 21-28.

Preiss, F. J. and J. A. Davidson. 1970. Characters for separating late-stage larvae, pupae, and adults of *Alphitobius diaperinus* and *A. laevigatus* (Coleoptera: Tenebrionicae) Annals. Entomol. Soc. Amer. 63: 807-808.

Propp, G and P. B. Morgan. 1985. Mortality of eggs and first-stage larvae of the house fly, *Musca domestica* L. (Diptera: Muscidae), in poultry manure. J. Kansas Entomol. Soc. 58: 442-447.

Rodriguez, J. G. and C. F. Wade. 1961. The nutrition of *Macrocheles muscaedomesticae* (Acarina: Macrochelidae) in relation to its predatory action on the house fly egg. Annals Entomol. Soc. Amer. 54: 782-788.

Rodriguez, J. G., C. F. Wade and C. W. Wells. 1962. Nematodes as natural food for *Macrocheles muscaedomesticae* (Acarina: Mesostigmata), a predator of the house fly egg. Annals Entomol. Soc. Amer. 55: 507-511.

Rodriguez, J. G., P. Singh and B. Taylor. 1970. Manure mites and their role in fly control. J. Med. Entomol. 7: 335-341.

Safrit, R. D. and R. C. Axtell. 1984. Evaluations of sampling methods for darkling beetles (*Alphitobius diaperinus*) in the litter of turkey and broiler houses. Poultry Sci. 63: 2368-2375.

Singh, P., W. E. King and J. G. Rodriguez. 1966. Biological control of muscids as influenced by host preference of *Macocheles muscaedomesticae*. J. Med. Entomol. 3: 78-81.

Smith, C. A. 1975. Observations on the life history, ecology and behavior of *Carcinops pumilio* (Erichson) M.S. thesis, University of New Hampshire.

Smith, R. 1981. Darkling beetles cause damage, nuisance, complaint. Feedstuffs (June): 13.

Stafford , K. C., III and D.E. Bay. 1987. Dispersion patterns and association of house fly, *Musca domestica* (Diptera: Muscidae), larvae and both sexes of *Macrocheles muscaedomesticae* (Acari: Macrochelidae) in response to poultry manure moisture, temperature, and accumulation. Environ. Entomol. 16: 159-164.

Stafford, K. C., III, C. H. Collison, J. G. Burg and J. A. Cloud. 1988. Distribution and monitoring lesser mealworms, hide beetles, and other fauna in high-rise, caged-layer poultry houses. J. Agric. Entomol. 5: 89-101.

Sukarsih, F., C. J. Geden, and R. C. Axtell. 1989. Filth fly oviposition and larval development in poultry manure of various moisture levels. J. Entomol. Sci. 24: 224-231.

Thornberry, F. D. 1978. Lesser mealworms invade Maine residences. Poultry Digest 37: 464.

Toyama, G. M. and J. K. Ikeda. 1976. An evaluation of fly predators at animal farms on leeward and central Oahu. Proc. Hawaii Entomol. Soc. 22: 369-379.

Turner, E. C., Jr. 1986. Structural and litter pests. Poultry Sci. 65: 644-648.

Vaughan, J. A., E. C. Turner, Jr. and P. L. Ruszler. 1984. Infestation and damage of poultry house insulation by the lesser mealworm, Alphitobius diaperinus (Panzer). Poultry Sci. 63: 1094-1100.

Wade, C. F. and J. G. Rodriguez. 1961. Life history of *Macrocheles muscaedomesticae* (Acarina: Macrochelidae), a predator of the house fly. Annals Entomol. Soc. Amer. 54: 776-781.

Wallace, M. M. H., R. G. Winks and J. Voestermans. 1985. The use of a beetle, *Alphitobius diaperinus* (Panzer) (Coleoptera: Tenebionidae) for the biological control of poultry dung in high-rise layer houses. J. Australian Inst. Agric. Sci. 51: 214-219.

Wenzel, R. L. 1955. The histerid beetles of New Caledonia (Coleoptera: Histeridae). Fieldiana: Zoology: 37: 601-637.

Willis, R. R. and R. C. Axtell. 1968. Mite predators of the house fly: a comparison of *Fuscuropoda vegetans* and *Macrocheles muscaedomesticae*. J. Econ. Entomol. 61: 1669-1674.

Wise, G. U., M. K. Hennessey and R. C. Axtell. 1988. A new species of manure-inhabiting mite in the genus *Poecilochirus* (Acari: Mesostigmata: Parasitidae) predacious on house fly eggs and larvae. Annals Entomol. Soc. Amer. 81: 209-224.

15. Mass Production of the Predaceous Mite, *Macrocheles muscaedomesticae* (Scopoli) (Acarina: Macrochelidae), a Predator of the House Fly

Chyi-Chen Ho, Harvey L. Cromroy and
Richard S. Patterson

ABSTRACTS

A mass-production method was developed for the predaceous mite, *Macrocheles muscaedomesticae* (Scopoli). Using spent house fly media, frozen house fly eggs, and a nematode, *Protorhabditis* sp., 2500 mites could be produced in eight days from 34.5 female mites at 30°C. Under these conditions, the generation time of this mite was 3.2 days and the mite population could increase as high as 69.8 times per generation. The relationship of the nematode to the rearing of *M. muscaedomesticae*, and the possible ways to improve the mass-production method are discussed.

INTRODUCTION

M. muscaedomesticae (Scopoli) is commonly found in chicken manure (Axtell 1963a; Filipponi 1955), and is the most abundant and, usually, the only macrochelid mite found in chicken manure (Axtell 1963a). This predaceous mite feeds on house fly eggs and first instar larvae. It has a reproductive rate, higher than the house fly and six other macrochelid mites (Cicolani 1979). Adult female mites are phoretic on house flies (Ewing 1913; Filipponi 1955; Hoffmann et al. 1974; Pereira and de Castro 1945; Rak 1972) and will disperse with house flies. *M. muscaedomesticae* has been shown to be a good agent to control house flies (Axtell 1963b, 1963c, 1966, 1968, 1970a, 1970b; Ito 1977a; Peck and Anderson 1970; Rodriguez et al. 1970; Singh et al. 1966; and Wicht and Rodriguez 1970).

The production of *M. muscaedomesticae* has been studied by Ito (1973, 1977b), Rodriguez and Wade (1961), Rodriguez et al.

(1962), and Wallwork and Rodriguez (1963). None of these were studied in large scale. This paper presents a primitive method to mass-produce *M. muscaedomesticae.*

MATERIALS AND METHODS

Adult female *M. muscaedomesticae* were collected in 1982 from caged-hen manure in poultry houses of the Department of Poultry Science, IFAS, University of Florida, Gainesville, Florida. An 80-ounce oblong food saver was used to culture the mites. A hole, 11.5 x 16.0 cm^2, was cut in the center of the lid and sealed with C-299 maiden chiffon for ventilation. Spent house fly media (Morgan 1981a) was used as a culturing medium. House fly eggs were supplied as the food. Both the spent house fly media and the house fly eggs were obtained from the "Insects Affecting Man and Animals Research Laboratory" of USDA in Gainesville, Florida. Fly eggs were chilled to prevent hatching. Spent house fly media was frozen before use for at least three days to kill unwanted organisms and defrosted to room temperature immediately before use.

Nematodes have been reported to be an excellent food for *M. muscaedomesticae* (Filipponi and di Delupis 1963; Ito 1971, 1977a; Rodriguez et al. 1962; Singh and Rodriguez 1966). A nematode, *Protorhabditis* sp. was found abundant in the stock colony, presumedly being brought in by *M. muscaedomesticae*, and multiplied excellently on the fly media. Widemouthed quart glass mason jars filled with fly media were used for the culture of the nematode. Under room temperature (21.1-26.7°C) the addition of one to two tablespoonfuls of the old nematode colony media to each mason jar produced great numbers of nematodes after three days. *Protorhabditis* sp. was supplied in the form of three to five day old nematode media. These mason jars were covered by the same C-299 maiden chiffon which was fastened with a rubber band. The maiden chiffon provides good ventilation and good isolation of arthropods.

I. Basic Tests

Four factors, the harvest time, the number of mites, the quantity of fly eggs, and nematode media to be introduced to each box (the 80-ounce food saver) were tested in sequence. Only adult female mites were used for introduction. Fly eggs were measured in ml following the method of Morgan (1981a). Nematode media was measured with a tri-pour disposable beaker. At the end of each test, the mites in each box were separated by a Tullgren Berlese funnel into 99.5% isopropyl alcohol. The method of Ing (1978) was followed to estimate the number of mites produced.

Harvest time. 400 grams of spent fly media, ten mites, one ml fly eggs, and two-thirds beaker nematode media were added to each box. The number of mites produced was separated and counted at the sixth, eighth, and tenth day. Twelve boxes were used for each treatment.

Number of *M. muscaedomesticae*. 600 grams of fly media, two ml fly eggs, two-thirds beaker nematode media were added to each box. Then, ten, twenty, thirty, or forty mites were introduced and allowed to propagate eight days. Ten boxes were used per treatment.

Quantity of house fly eggs. 600 grams of fly media, twenty mites and one beaker nematode media were added to each box. The testing levels of fly eggs were zero, two, four and six ml. The mites were allowed to propagate eight days. Each treatment consisted of ten boxes.

Quantity of nematode media. 600 grams of fly media, twenty mites, and two ml fly eggs were added to each box. One, one-and-a-half, or two beakers of fly eggs were tested. The mites were separated after eight days and counted. Six boxes were used for each treatment.

II. Mass production tests

To simplify this process, a few changes were made after these basic tests. First, a 32-ounce oblong food saver was used to measure the fly media. To each 80-ounce box, two, 32-ounce boxes of fly media were added. Second, a meatballer scoop (6.8 cm^3 in volume) was used to introduce the mites. Third, fly eggs did not appear to be necessary for mass-producing the mites.

Number of mites per meatballer scoop. Ten samples were randomly ladled from each of two boxes of seven-day-old stock colony. The media in the box was stirred before ladling. Female deutonymphs were also distinguished and recorded.

Mites produced by mass-production method. The number of mites produced per box by meatballer method was tested. One scoop of eight-day-old stock colony media was ladled to a new box. As these media contained nematodes, one or zero beakers of nematode media was added to each box. The mites were separated after eight days. Eight boxes were maintained for each treatment. All of the above tests were conducted at $30\pm2°C$, 60-98% R.H., and 14:10 (L:D).

Generation Time. Newly emerged adult females (within six hours) and males were transferred to one-ounce plastic cups that contained fly media and nematode media in a 2:1 ratio, one pair per cup. These cups were examined at six-hour intervals, ten cups at each time, for the presence of a second adult female mite -- the daughter mite. The experiment was terminated when the daughter mite(s) was found in each of the ten examined cups. Four temperatures, 22, 26, 30, and 34°C, were studied.

RESULTS AND DISCUSSION

Studies of Cicolani (1979), Filipponi (1971), Filipponi and Petrelli (1967), and Filipponi et al. (1971) showed that the optimum temperature for population increase of this mite is 30°C. Therefore, 30°C was chosen for mass producing the mites. Based on the study of Rodriguez and Wade (1961), as well as on consideration of the cost of production, fly media was chosen for mass production of *M. muscaedomesticae*. The study of Rodriguez and Wade (1961) showed the addition of soybean oil meal and fish meal could increase the number of progeny of this mite. In preliminary tests, the addition of fish emulsion fertilizer and commercial layer feed did not produce more mites. Therefore, these nitrogenous food supplements were not used.

I. Basic Tests

The results of the four basic tests are shown in Table 1. The test on the harvest time showed that eight days is the proper culture time. As the 80-ounce box could hold more media, 600 gm of fly media were used in the following tests. The introduction of twenty mites gave significantly more progeny than the introduction of ten mites. Introducing more than twenty mites did not produce more progeny, and a crowding effect was observed as indicated by the decrease in the number of mites produced. The supply of fly eggs had no effect on the number of mites produced. This differed from

Table 1. Population of _M. muscaedomesticae_ (Mean±SE) as affected by various factors at 30+2±°C, 60-98% R.H.

Factor tested	Testing levels			
	No. mites produced			
Harvest time (xth day)	6	8	10	
	963.3c±128.2	3706.7a±322.4	2498.6b±346.6	
No. adult females introduced	10	20	30	40
	2390a±245.2	3664b±317.6	3146ab±223.0	3086ab±291.4
Quantity of house fly eggs (ml)	0	2	4	6
	2846a±257.4	2854a±173.5	2846a±117.3	2619a±245.7
Quantity of nematode medium (Beaker)	1.0	1.5	2.0	
	4417a±365.3	4247a±309.1	3473a±296.7	

Mites were separated after eight days except the test of harvest time.
Means in the same row that followed by the same letter were not significantly different at the 0.05 level in DMRT.

the results of Rodriguez et al. (1962) and Ito (1973). However, fly eggs were still offered.

Three tested volumes of nematode media all gave better result than previous tests. However, there was no significant difference among these three volumes. The number of mites produced seemed to decrease with an increase in the nematode media. Propagation of nematodes requires wetting the fly media, which makes it unsuitable for production of mites. Rodriguez et al. (1962) "pre-seeded" the media with nematodes and produced lower numbers of mites. That may be caused by the same reason. Therefore, one beaker of nematode media was chosen for the mass production of the mites.

The excellent food value of the nematode to this mite was further substantiated. _Protorhabditis_ sp. is a new species of nematode found to be prey of *M. muscaedomesticae*. Poinar (1965) reported individuals of *M. glaber* and *M. submotus* carried this nematode. *Protorhabditis* sp. was observed to attach to the coxal cavity, gnathosomal cavity, peritremes, and wrinkles on the body surface of the mites. Both male and female mites can carry nematodes. Among the observed mites, 100% of starved mites carried the nematodes. However, only a few well-fed mites carried nematodes. The rhabditid nematodes almost invariably carry with them the bacteria that they feed on. Consequently, the macrochelid mites will not have problems with an adequate alternative food supply as generally happens with predators.

Mass production of the nematodes had been studied by Singer and Krantz (1967) and Singh et al. (1966). The method described here is more economic and convenient.

II. Mass production tests

Through the basic tests, a primitive rearing method was developed. Using the 80-ounce plastic box, 600 gm of fly media, and one beaker of nematode media, over 2800 mites can be produced from twenty adult female mites in eight days. Furthermore, house fly eggs can be omitted as a food.

However, it is not realistic, in a mass-production technique, to weigh medium and to count the mites for introduction. To simplify the procedure, the 32-ounce box was chosen to measure the fly media. Two measures of fly media with the 32-ounce box well filled the 80-ounce box. A meatballer scoop was selected to introduce mites from old/stock colony to a new box.

A test was conducted to determine the number of mites that were ladled per meatballer scoop. The results are shown in Table 2. The sex of the mature deutonymph of this mite is distinguishable. At 30°C, the deutonymphs will develop to the adult stage within half a day to one day. Therefore, the number of female deutonymphs was recorded and considered as adults. The distribution of this mite in the meatballer samples was uniform, with standard errors less than 10% of the mean. The sum of female adults and female deutonymphs, 34.5, is slightly higher but still acceptable (see Table 1). Samples of a smaller quantity had a SE/mean ratio larger than 0.10. Hence, the meatballer method was considered acceptable for introducing mites. The presence of male mites increased the propagation of the mite.

Table 2. Number of *M. muscaedomesticae* in one scoopful of eight-day-old stock colony media.

	Female	DN*	Subtotal	Male	Nymph	Total
Mean**	24.65	8.85	34.50	14.70	19.65	68.85
SE	2.27	0.77	2.43	1.17	1.40	4.17
Mean/SE	0.092	0.087	0.094	0.079	0.071	0.061

*The mature deutonymph which will soon emerge to adult female.
**Mean of 20 scoops.

After the selection of the meatballer scoop and the 32-ounce box, the mite numbers produced with this method were tested. The material ladled by the meatballer scoop contained not only mites but also nematodes. Would these nematodes be sufficient to multiply and serve as food for the mite population? This was also tested and the results are shown in Table 3. Compared with previous tests which introduced twenty adult female mites, the mite population produced with this method was significantly smaller than those tests on the quantity of nematode media and the number of mites but not significantly different to the test on the quantity of fly eggs. However, without additional nematodes, the mite population produced was significantly smaller. Therefore, it was necessary to add one beaker of nematode media.

Because adult females not only eat much more than the other stages of this mite but also proliferate the population, it is obviously more important to produce female mites. Therefore, the sex ratio

Table 3. *M. muscaedomesticae* produced with the meatballer scoop method at 30±2°C, 60-98% R.H.

Quantity of nematode medium (beaker)	No. mites after 8 days		SE/Mean
	Mean	SE	
1	2270.0a	77.6	0.038
2	1755.5b	89.9	0.051

Means followed by the same letter were not significantly different at the 0.05 level in DMRT.

(females/total adults) at different culture times was studied. The results are shown in Table 4. The proportion of the adult female mites (the sex ratio and the ratio of adult females to total population) increased as the colony aged. The sex ratio of this mite was subjected to change with the environmental conditions (Filipponi and Petrelli 1967; Filipponi et al. 1971; and Ito 1977b). There would be only adult female mites after eleven to twelve days. However, the increase of the proportion of the adult female mites resulted from the decrease in the number of males, young, and total population numbers. The study of Ito (1977b) also showed this.

The multiplication of the nematodes caused wetting and destruction of the substratum, making it unsuitable for both the nematodes and the mites. After the eighth day, the quality of the substratum and the quantity of food all decreased. Because of this and also sanitary considerations, eight days was selected as the best culture time.

Feeding on house fly eggs, the generation time (egg to egg) of this mite was 5.1 days and the fecundity was approximately 125 eggs per female, at $30\pm2°C$ (Ho 1985). The color and the shape of fly media made it too difficult to study the life history of M. *muscaedomesticae* under mass production conditions. But the generation time was studied. The generation time (adult female to adult female) was 5.7 ± 0.1, 3.3 ± 0.1, 3.2 ± 0.1, and 3.7 ± 0.1 days at 22, 26, 30, and 34°C, respectively. It was shortest at 30°C, and shorter than the data of Cicolani (1979) and Ho (1985). The Ro of this mite in fly media and while feeding on nematodes was estimated to be 69.85 (Ho 1985), indicating that this mass production method

Table 4. The proportion of the adult female mites in the mass-production population of M. *muscaedomesticae* at different days at $30\pm2°C$, 60-98% R.H.

Days after set up	females/total adults	females/total population
6	0.5355a	0.4298a
7	0.5853ab	0.5015ab
8	0.6252b	0.5324b
9	0.7317c	0.6661c

See Table 3.

favored the propagation of *M. muscaedomesticae*. Based on the generation time, 30°C was determined to be the optimal temperature for the mass production of this mite.

Through the above studies, a mass-production method for *M. muscaedomesticae* was developed. In summation, it is as follows:

Mite colony.
Container: an 80-ounce oblong food saver.
Substratum:
  1. Spent house fly media: two boxes, measured by a 32-ounce oblong food saver.
  2. Nematode media: one beaker, three- to five-day-old, measured by a fifty ml tri-pour disposable beaker.
Quantity of *M. muscaedomesticae* to introduce: one ladle of eight-day-old colony media, measured by a meatballer of $6.8cm^3$ in volume.
Nematode colony.
Container: widemouthed quart glass mason jar.
Substratum: spent house fly media, filled up to four cm beneath the mouth.
Quantity of nematodes to introduce: one to two tablespoonfuls of three- to five-day-old nematode colony.
Culture time: three to five days.

For quality control, two colonies should be maintained. One as a stock colony in which house fly eggs are offered to prevent a change in food habits of the mites. The other as a "propagating colony," in which house fly eggs are not offered for field releases. Adult males should be collected from the field and added to the stock colony to continue to improve the gene pool.

Since the developed mass-production method for *M. muscaedomesticae* uses spent house fly media and does not require many house fly eggs, it would be economical and profitable to include mass production of the parasites of the house fly, e.g. *Spalangia endius*. Referring to the report of Morgan (1981a, b), approximately 479,768 *Spalangia endius* can be produced together with every 500,000 *M. muscaedomesticae* for field release, assuming 2500 mites are produced per box. This would greatly benefit the biological control of house flies.

Future improvement of this method

Over 4000 mites were produced per box; however, the majority of tests produced less than 3000 mites per box. Studies showed that a production of 3000 to 4000 mites per box was reasonable. Both the mite numbers produced and the cost could be improved. These may be accomplished in the following manner:

1. To control the quality of the spent house fly media, especially the water content. The quality of fly media varied from time to time. This affected the multiplication of the nematodes and the mites, and might contribute to the variation of the mite numbers produced.

2. To control and maintain the relative humidity in the range of 70-80%.

3. To maintain good ventilation to prevent ammonia concentration in the air. Ammonia has an adverse effect on the reproduction of this mite (Wallwork and Rodriguez 1963).

4. To maintain good sanitation to prevent the occurrence of small ascid mites, which could destroy both the mite and the nematode colony. Every mite colony box should be cleaned and sterilized after each use.

5. To reduce the number of *M. muscaedomesticae* ladled per scoop.

6. To reduce the quantity of nematode media introduced.

7. To use fresh house fly media instead of using spent house fly media. This could make the mass production of these mites independent to the production of house flies.

REFERENCES CITED

Axtell, R. C. 1963a. Acarina occurring in domestic animal manure. Ann. Entomol. Soc. Am. 56:628-633.

_____. 1963b. Effect of Macrochelidae (Acarina: Mesostigmata) on house fly production from dairy cattle manure. J. Econ. Entomol. 56:317-321.

_____. 1963c. Manure-inhabiting Macrochelidae (Acarina: Mesostigmata) predaceous on the house fly. Advances in Acarology 1:55-59.

_____. 1966. Comparative toxicities of insecticides to house fly larvae and *Macrocheles muscaedomesticae*, a mite predator of the house fly. J. Econ. Entomol. 59:1128-1130.

Axtell, R. C.. 1968. Integrated house fly control: Populations of fly larvae and predaceous mites, *Macrocheles muscaedomesticae,* in poultry manure after larvicide treatment. J. Econ. Entomol. 61:245-249.

_____. 1970a. Integrated fly control program for caged-poultry houses. J. Econ. Entomol. 63:400-405.

_____. 1970b. Fly control in caged-poultry houses: Comparison of larviciding and integrated control programs. J. Econ. Entomol. 63:17341737.

Cicolani, B. 1979. The intrinsic rate of natural increase in dung macrochelid mites, predators of *Musca domestica* eggs. Boll. Zool. 46:171-178.

Ewing, H. E. 1913. A new parasite of the house fly (Acarina, Gamasoidea). Entomol. News 24:452-456.

Filipponi, A. 1955. Sulla natura dell'associazione tra *Macrocheles muscaedomesticae e Musca domestica.* Riv. Parassitol. 16:83-102.

_____. 1964. The feasibility of mass producing macrochelid mites for field trials against house flies. Bull. Wld. Hlth. Org. 31:499-501.

_____. 1971. Influence of temperature on population increase of 4 dung living macrochelid mites within optimum range. Pages 775-779 in Proc. 3rd. Intl. Cong. Acarol., Prague.

Filipponi, A., and G. D. di Delupis. 1963. Sul regime dietetico di alcuni Macrochelidi (Acari: Mesostigmata), associati in natura a muscidi di interesse sanitario. Riv. Parassitol. 24:277-288.

Filipponi, A., B. Mosna, and G. Petrelli. 1971. The optimum temperature of *Macrocheles muscaedomesticae* as a population attribute. Riv. Parassitol. 32:193-218.

Filipponi, A., and G. Petrelli. 1967. Autoecology and capacity for increase in numbers of *Macrocheles muscaedomesticae* (Scopoli)(Acari: Mesostigmata). Riv. Parassitol. 28:129-156.

Ho, C. C. 1985. Mass production of the predaceous mite, *Macrocheles muscaedomesticae* (Scopoli)(Acarina: Macrochelidae), and its potential use as a biological control agent of house fly, *Musca domestica* L. (Diptera: Muscidae). Ph. D. Dissertation, Univ. of Florida, Gainesville. 186pp.

Hoffmann, A., I. B. de Barrera, and C. Mendez. 1974. New records of mites from Mexico. Revista de la Sociedad Mexicana de Historia Natural 33:151-159.

Ing, R. T. H. 1978. Seasonal variations of mites of the suborder Mesostigmata (Acarina) from south Florida turfgrasses. Ph. D. Dissertation, Univ. of Florida, Gainesville. 122pp.

212

Ito, Y. 1973. The effects of nematode feeding on the predatory efficiency for house fly eggs and reproduction rate of *Macrocheles muscaedomesticae* (Acarina: Mesostigmata). Jap. J. Sanit. Zool. 23:209-213.

_____. 1977a. Predatory activity of mesostigmatid mites (Acarina: Mesostigmata) for house fly eggs and larvae under feeding nematodes. Jap. J. Sanit. Zool. 28:167-173.

_____. 1977b. Changes of the population density and stage composition of three manure inhabiting mesostigmatid mite species (*Macrocheles muscaedomesticae, Parasitus gregarius, Uroobovella marginata*) on a restricted food supply. Jap. J. Appl. Entomol. Zool. 21:74-78.

Morgan, P. B. 1981a. Mass production of *Musca domestica* L. Pages 181-191 in Proc. of Workshop "Status of Biological Control of Filth Flies." SEA, USDA, Gainesville, Florida.

_____. 1981b. Mass production of *Spalangia endius* Walker for augmantative and/or inoculative field releases. Pages 185-188 in Proc. of Workshop "Status of Biological Control of Filth Flies." SEA, USDA, Gainesville, Florida.

Peck, J. H., and J. R. Anderson. 1970. Influence of poultry manure removal schedule on various Diptera larvae and selected arthropod predators. J. Econ. Entomol. 63:82-90.

Pereira, C., and M. P. de Castro. 1945. Contribution to the knowledge of the type species of *Macarocheles* Latr. (Acarina): *M. muscaedomesticae* (Scopoli, 1772) Emend. Arq. Inst. Biol. 16:153-186.

Poinar, G. O. 1965. An association between *Pelodera (Coarctadera) acarambates* n. sp. (Rhabditina: Nematoda) and macrochelid mites *Macrocheles glaber* and *M. submotus* Mesostigmata: Acari). Nematologia 10:507-511.

Rak, H. 1972. Incidence of *Macrocheles muscaedomesticae* (Scopoli) (Acarina: Macrochelidae): The external parasites of Musca domestica and related flies in Iran. Tehran Univ. Vet. Fac. J. 27:1-9.

Rodriguez, J. G., P. Singh, and B. Taylor. 1970. Manure mites and their role in fly control. J. Med. Entomol. 7:335-341.

Rodriguez, J. G., and C. F. Wade. 1961. The nutrition of *Macrocheles muscaedomesticae* (Acarina: Macrochelidae), a predator of the house fly egg. Ann. Entomol. Soc. Am. 54:782-788.

Rodriguez, J. G., C. F. Wade, and C. N. Wells. 1962. Nematodes as a natural food for *Macrocheles muscaedomesticae* (Acarina: Macrochelidae) a predator of the house fly egg. Ann. Entomol. Soc. Am. 55:507-511

Singer, G., and G. W. Krantz. 1967. The use of nematodes and oligochaetes for rearing predatory mites. Acarologia 9:485-487.

Singh, P., W. E. King, and J. G. Rodriguez. 1966. Biological control of muscids as influenced by host preference of *Macrocheles muscaedomesticae* (Acarina: Macrochelidae). J. Med. Entomol. 3:78-81.

Singh, P., and J. G. Rodriguez. 1966. Food for macrochelid mites (Acarina) by an improved method for mass rearing of a nematode, *Rhabditella leptura*. Acarologia 8:549-550.

Wallwork, J. H., and J. G. Rodriguez. 1963. The effect of ammonia on the predation rate of *Macrocheles muscaedomesticae* (Acarina: Macrochelidae) on house fly eggs. Advances in Acarology 1:60-69.

Wicht, M. C., Jr., and J. G. Rodriguez. 1970. Integrated control of muscid flies in poultry houses using predator-mites, selected pesticides and microbial agents. J. Med. Entomol. 7:687-692.

16. The Effect of *Bacillus thuringiensis var. thuringiensis* on *Musca domestica* L. Larvae Resistant to Insecticides

J. B. Jespersen and J. Keiding

INTRODUCTION

Bacillus thuringiensis strains produce two main types of toxins, delta-endotoxins and beta-exotoxins, which are toxic to many kinds of insects. The endotoxin is a protein crystal produced at sporulation, while the exotoxin is an adenine nucleotid with glucose produced in the growth phase. The endotoxin is mainly used for control of various lepidopterous insect pests, while the exotoxin is effective against species of Diptera, Hymenoptera, Coleoptera and Orthoptera (Carlberg and Lindström 1987). The exotoxin is toxic to fly larvae because it affects moulting and pupation (Carlberg et al. 1985).

From 1983-1986 preparations with *Bacillus thuringiensis var. thuringiensis* containing beta-exotoxin were tested by the Danish Pest Infestation Laboratory (DPIL) for the effect on larvae of the house fly, *Musca domestica*, to evaluate the possibility of using the preparation for fly control on farms. The idea was to apply the formulation to the surface of the larval biotope; the bacteria should then survive, the spores germinate and reproduce and produce more exotoxin to kill the fly larvae present.

As insecticide resistance is a very serious problem in house fly control, investigations were carried out on the possible cross-resistance between the contact effect of insecticides and larvicidal effect of *Bacillus thuringiensis* exotoxin in house fly strains representing a wide range of resistance to OP's and pyrethroids. This paper is mainly focused on these cross-resistance investigations.

215

MATERIALS AND METHODS

Materials

Formulation of *Bacillus thuringiensis var. thuringiensis* (or B.th.-Hl). The formulation tested was named Muscabac and was produced in Finland by Farmos Group Ltd. Muscabac is a bacterial culture, having exotoxin and spores as active components. Muscabac was registered in Finland in 1981 for fly control in piggeries, hen houses, and compost toilets. It was stated by the company to contain approximately 200 mg exotoxin per litre.

House fly Strains. In the laboratory tests reported fourteen strains of house flies *Musca domestica*, seven Danish field strains and seven laboratory strains, were used. The field strains were newly collected on farms where trials of *Bacillus* formulation were carried out. These strains were representative of the insecticide resistance pattern of house flies on Danish farms with a long history of insecticide use for fly control and development of multi-resistance (Keiding 1977, Keiding and Jespersen 1986).

The laboratory strains represented susceptibility as well as different patterns and levels of resistance to insecticides:

WHO-SRS: A susceptible standard reference strain obtained from the University of Pavia.

$40j_2$: A strain collected on a Danish farm in 1980 and kept without insecticide pressure at the DPIL. It is used at DPIL for testing the efficacy of insecticides and formulations.

213ab w4: A white-eyed mutant of a pyrethroid-resistant strain collected in Sweden 1957. The strain kept without selection pressure and the original pyrethroid and OP-resistance reverted to susceptibility.

Zürich-b: A multi-resistant strain collected at a Swiss farm in 1977. Occasional laboratory selection with permethrin since 1979. High resistance to pyrethroids (super-kdr), low to OP's.

381zb: A multi-resistant strain collected on a Danish farm in 1978. Selection 1-2 times a year with permethrin from 1979 and also with dimethoate from 1983. High resistance to OP's and pyrethroids (super-kdr).

571ab: Collected 1980 at a garbage dump at Tokyo, Japan. Very high resistance to most OP's, susceptible to pyrethroids. Two selections with fenitrothion in 1983 before the tests with *Bacillus* exotoxin.

645ab: A multi-resistant strain collected on a Swedish farm 1982. Laboratory-pressure with trichlorfon feeding.

More information on the insecticide resistance of the fly strains is given in Tables 2a and 2b.

Test Methods

Larvicide tests with artificial larval breeding medium were carried out in small 100 ml plastic cups with 20 grams of larval medium. The artificial medium consisted of a mixture of 400 grams wheat bran, 200 grams alfalfa meal, 10 grams baker's yeast, 15 grams malt extract, 500 grams milk, and 500 ml water.

To each 20 gram medium 2.5 ml of *Bacillus* formulation diluted with water was added. One to two hours later the medium was seeded either with 1st instar larvae hatched from twenty-five eggs placed on wet filter paper in a beer cap on top of the medium (Figure 1) or with twenty-five early 3rd instar (2.5 days old) larvae placed on the medium with tweezers. The number of adult flies emerged was counted after three to four weeks. The tests were made at 27°C and 70-80% humidity.

The tests with manure as a larval medium were made in a slightly different way. The containers were 240 ml plastic cups with 80 grams of manure in each cup (Figure 2); 10 ml of diluted *Bacillus* formulation was applied. Each cup was confined in a cage in which emerging flies and larvae escaping from the medium were collected for counting. Eggs or 3rd instar larvae were added as in tests with artificial media.

The containers were placed at 22°C and 70% RH. Hen and calf manure was used in these trials. The manure was stored frozen until one day before use.

Three replicates of ten dilutions of the *Bacillus* formulation were used per fly strain and type of medium, while six replicates were used for the control with plain water.

On the basis of larval plus pupal mortality, corrected for control mortality by Abbot's formula, the LC_{50} and LC_{95} were determined by the probit method (Finney 1971).

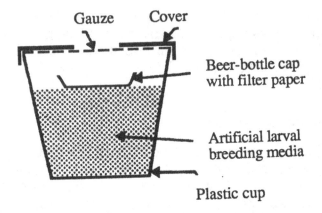

Figure 1. The figure shows how the larvicide tests were carried out in artificial larval breeding media. For explanation, see the text.

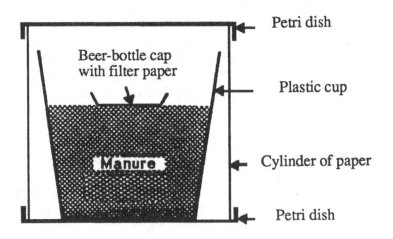

Figure 2. The figure shows how larvicide tests were carried out with hen and calf manure. For explanation see the text.

RESULTS

<u>Effect in Different Larval Media</u>

    A comparison of the effect of a liquid formulation of B.th. H-1 in laboratory tests with artificial medium, calf manure and hen manure is shown in Table 1. In tests with artificial medium and calf manure started with newly emerged larvae the LC95 was below 2ppm exotoxin, but in hen manure the LC was about three times higher. In tests started with third instar larvae there was also less effect in hen manure. Table 1 shows that the LC95s were about three times higher than the LC50s, and the lethal concentrations were two to five times higher when tests were started with third instar larvae than in tests with newly emerged larvae.

<u>Effect in Fly Strains Compared to Insecticide Resistance</u>

    The LCs of exotoxin were estimated in tests with newly emerged and with third instar larvae in seven field strains collected on farms where B.th. formulations were tried, and in seven laboratory strains with different patterns and levels of resistance to insecticides. All the results of tests in artificial medium are shown in Tables 2a and 2b, but in the following only results showing linearity

Table 1. Mean lethal concentrations of exotoxin in artificial medium, calf and hen manure in tests with a liquid formulation of *B. th.* H-1 with several strains of *Musca domestica* (cf. Table 2 and 3). Only results showing linearity in the probit analysis have been used for calculating the means.

| | 1st instar larvae | | 3rd instar larvae | |
| | LC50 | LC95 | LC50 | LC95 |
	(ppm exotoxin)		(ppm exotoxin)	
Artificial medium	0.6	1.6	2.9	7.3
Calf manure	0.7	1.7	1.3	3.8
Hen manure	2.0	6.0	3.0	11.0

Table 2a. Resistance to insecticides and larvicidal effect of *Bacillus thuringiensis* var. *thuringiensis* liquid formulation in artificial medium in tests with several strains of *Musca domestica*.

Resistance to insecticides determined by topical application tests.
S = susceptible; L = low; M = moderate; H = high; VH = very high resistance
R/S = resistance ratio compared to standard reference strain.
f.l. = fiducial limits of LC; SRS = standard susceptible reference strain.

Strain / Field strain / Farm Collection	Resistance to (topical application)						Larvicidal effect: ppm exotoxin			
	Organo-phos-phorous	Dimethoate R/S at LD$_{50}$	LD$_{95}$	Pyre-throid	BRM/PB R/S at LD$_{50}$	LD$_{95}$	First instar larvae LC$_{50}$	(95% f.l.)	LC$_{95}$	(95% f.l.)
268 z2	M-H						0.6	(0.4-0.7)	1.6	(1.3-2.2)
666 a	M-H	26	38	M-H	5	25	0.8	(0.6-1.0)	2.2	(1.5-7.2)
560 a	M-H	14	37	M-H	5	43	0.6	(0.4-0.7)	1.3	(1.0-1.9)
657 c							0.8	(0.3-1.3)	1.8	(1.2-12.8)*
– d		24	89	M	4	14				
– e	M-H									
588 d	M-H	13	41	L	2	2	0.8	(0.6-0.9)	2.1	(1.7-2.8)
– e										
606 e		22	36	M-H	23	19	1.1	(0.0-1.9)	3.4	(1.9-10.8)*
– f	M-H			M-H			0.1		0.5	*
– g										

(Continued)

Table 2a.1 (Cont.). Resistance to insecticides and larvicidal effect of *Bacillus thuringiensis var. thuringiensis* liquid formulation in artificial medium in tests with several strains of *Musca domestica*.

Resistance to insecticides determined by topical application tests.

S = susceptible; L = low; M = moderate; H = high; VH = very high resistance

R/S = resistance ratio compared to standard reference values; f.l. = fiducial limits of LC; SRS = standard susceptible reference strain.

| Strain | Resistance to (topical application) | | | | | | Larvicidal effect: ppm exotoxin | | | |
| | Organo-phosphorous | Dimethoate R/S at | | Pyrethroid | BRM/PB R/S at | | First instar larvae | | | |
Lab Strain		LD50	LD95		LD50	LD95	LC50	(95% f.l.)	LC95	(95% f.l.)
WHO-SRS	S	1.0	1.0	S-L	1.0	1.0	0.5	(0.5-0.6)	0.6	(0.6-0.6)
40j2	L-M			L	2.0	4.0	0.9	(0.7-1.1)	2.2	(1.8-3.0)
213abW4	S	1.1	1.3	S	1.1	1.1	0.2	(0.1-0.4)	1.1	(0.7-1.7)
Zürich-b	L	3.0	3.0	H-VH	56.0	68.0	1.2	(0.9-1.4)	3.0	(2.4-4.8)
381zb	H	25.0	47.0	H-VH	79.0	77.0				
571ab	M-VH			S			0.1	(0.0-0.2)	0.5	(0.3-0.7)
645ab	M-H	9.0	16.0[1,2]	M-H	7.0	11.0[3]	0.4	(0.3-0.5)	1.7	(1.2-2.9)

1) Dimethoate M. Fenitrothion, jodfenphos, fenthion etc. VH.

2) Other OP' H.

3) Permethrin H. R/S at LD 95 > 50.

* Significant deviation from linearity in the probit analysis.

Table 2b. Resistance to insecticides and larvicidal effect of *Bacillus thuringiensis var. thuringiensis* liquid formulation in artificial medium in tests with several strains of *Musca domestica*.

Resistance to insecticides determined by topical application tests.
S = susceptible; L = low; M = moderate; H = high; VH = very high resistance
R/S = resistance ratio compared to standard reference values; f.l. = fiducial limits of LC; SRS = standard susceptible reference strain.

Strain Field strain	Farm Collection	Resistance to (topical application) Organo-phosphorous	Dimethoate R/S at LD50	LD95	Pyre-throid	BRM/PB R/S at LD50	LD95	Larvicidal effect: ppm exotoxin Third instar larvae LC50	(95% f.l.)	LC95	(95% f.l.)
268	z2	M-H						2.6	(2.3-3.0)	5.5	(4.7-7.2)
666	a	M-H	26	38	M-H	5	25	5.1	(4.3-6.1)	17.4	(12.3-34.6)
560	a	M-H	14	37	M-H	5	43	3.4	(1.7-5.1)	6.4	(4.5-33.7)
657	c							3.4	(2.9-3.9)	8.5	(7.1-11.0)
–	d				M	4	14				
–	e	M-H	24	89							
588	d	M-H	13	41	L	2	2				
–	e							4.4	(2.8-5.3)	9.1	(7.2-17.9)
606	e				M-H	23	19				
–	f	M-H	22	36				2.6	(1.0-2.3)	36.2	*
–	g				M-H			1.7		5.5	(4.3-8.8)

(Continued)

Table 2b.1 (Cont.). **Resistance to insecticides and larvicidal effect of *Bacillus thuringiensis var. thuringiensis* liquid formulation in tests with several strains of *Musca domestica*.**

Resistance to insecticides determined by topical application tests.
S = susceptible; L = low; M = moderate; H = high; VH = very high resistance
R/S = resistance ratio compared to standard reference values; f.l. = fiducial limits of LC; SRS = standard susceptible reference strain.

Strain	Resistance to (topical application)						Larvicidal effect: ppm exotoxin Third instar larvae			
	Organo-phos-phorous	Dimethoate R/S at		Pyre-throid	BRM/PB R/S at					
Lab Strain		LD50	LD95		LD50	LD95	LC50	(95% f.l.)	LC95	(95% f.l.)
WHO-SRS	S	1.0	1.0	S-L	1.0	1.0	4.2	(3.4-4.9)	7.8	(6.6-11.3)
40j2	L-M			L	2.0	4.0	1.2	(1.0-1.4)	3.8	(3.1-5.0)
213abW4	S	1.1	1.3	S	1.1	1.1	1.8	(1.5-2.0)	4.2	(3.5-5.4)
Zürich-b	L	3.0	3.0	H-VH	56.0	68.0	3.2	(2.7-3.7)	9.2	(7.5-12.3)
381zb	H	25.0	47.0	H-VH	79.0	77.0	0.9	(0.6-1.2)	3.1	(2.5-4.5)
571ab	M-VH			S						
645ab	M-H	9.0	16.0[1,2]	M-H	7.0	11.0[3]				

1) Dimethoate M. Fenitrothion, jodfenphos, fenthion etc. VH.
2) Other OP' H.
3) Permethrin H. R/S at LD 95 > 50.
* Significant deviation from linearity in the probit analysis.

in the probit analysis are mentioned. In tests started with newly emerged (first instar) larvae the LC50 varied from 0.1 to 1.2 ppm and the LC95 from 0.5 to 3.0 ppm. In tests started with 2.5 day old third instar larvae the range of LC50 was 0.9 to 5.1 ppm and the LC95 varied from 3.1 to 17.4. There was a considerable variation between and within strains, but no clear relation between the larvicidal LC to *B. th.* exotoxin and resistance to the contact effect of insecticide in adults. The Japanese strain 571ab showing a very high resistance to OP's was most susceptible to exotoxin. The two laboratory strains, Zürich-b and 381zb (only 1st instar larvae were tested), with the highest resistance to pyrethroids showed a relatively high tolerance to B.th. exotoxin, but so did the unselected strain 40j2 which is fairly susceptible to pyrethroids.

Results of tests in calf manure with seven laboratory strains are shown in Table 3. They show less differences in the effect of exotoxin between the strains and no evidence of relation between resistance to insecticides and effect of exotoxin.

DISCUSSION

The results reported in this paper show no indication of cross-resistance between the contact effect of insecticides and the larvicidal effect of *B. th.* exotoxin in house fly strains with a great range of resistance to OP's and pyrethroids. This agrees with Rupes et al. (1987) who stated that there was no cross-resistance between beta-exotoxin and insecticides in six multiresistant field strains in Czechoslovakia, but they did not give figures on susceptibility tests with these strains.

Very little additional information on the effect of *B. th.* exotoxin in relation to insecticide resistance in insect strains is available. Carlberg and Lindström (1987) compared the effect of a formulation of *B. th.*-Hl with exotoxin as the active ingredient on two strains of *Drosophila melanogaster*, the Hikone-R strain resistant to lindane, DDT, parathion and nicotine and a normal susceptible strain. The resistant strain showed only 1.1-1.9 times higher LC of B.th., and the difference was not significant.

McGaughey & Johnson (1987) found that strains of the Indian meal moth, *Plodia interpunctella* made resistant to various strains of *B. th.* producing endotoxin were susceptible to exotoxin. Thus, so far there is no evidence of clear cross-resistance between traditional insecticides and *B. th.* exotoxin.

Table 3. Larvicidal effect of *Bacillus thuringiensis var. thuringiensis* liquid formulation in calf manure in tests with seven laboratory strains of *Musca domestica*.

Concerning the resistance to insecticides of these strains see Table 2.

Lab strain	Larvicidal effect: ppm exotoxin							
	First instar larvae				Third instar larvae			
	LC50	(95% f.l.)	LC95	(95% f.l.)	LC50	(95% f.l.)	LC95	(95% f.l.)
WHO - SRS	0.6	(0.2-0.8)	1.7	(1.2-4.1)	0.9	(0.8-1.0)	1.6	(1.4-2.1)
40j2	1.5	(0.9-1.9)	3.0	(2.3-7.1)	1.5	(1.2-1.8)	3.9	(3.2-5.4)
213abw4	0.4	(0.3-0.5)	0.8	(0.7-1.2)	1.3	(1.0-1.5)	4.5	(3.6-6.2)
Zürich-b					1.5	(1.1-1.9)	5.1	(3.9-7.4)
381zb	1.4	(0.5-2.2)	3.5	(2.2-2.3)*	1.9	(1.3-2.4)	3.9	(2.9-8.2)*
571ab	0.7	(0.0-1.0)	1.9	(1.4-)	1.7	(0.0-3.6)	9.7	(4.7-14.3)*
645ab	0.2	(0.1-0.2)	1.0	(0.7-1.6)	2.0	(1.8-2.2)	2.3	(2.1-2.6)*

* Significant deviation from linearity in the probit analysis

However, strains of *Musca domestica* may develop moderate (six to fourteen fold) resistance to exotoxin when exposed to selection with formulations of *B. th.*-Hl formulations (Harvey & Howell 1965 and Wilson & Burns 1968), and moderate resistance has also been induced in a strain of *Drosophila melanogaster* (Carlberg & Lindström 1981, 1987).

Similarly, no significant cross-resistance to the effect of *B. th.* strains producing endotoxin has been found in various strains of Lepidoptera: *Plutella, Spodoptera* and *Heliothis* (Iman et al. 1986, Yeh et al. 1986, Ignoffo & Roush 1986), and *Diptera: Anopheles, Culex* and *Simulium* (Sun et al. 1980, Yang et al. 1985, Guillet 1984), but selection pressure with such *B. th.* strains may develop significant resistance quite rapidly in species of moths (McGaughey 1985, McGaughey & Beeman 1988, Stone et al. 1989) and some increase of tolerance in strains of mosquitoes (Georghiou & Vasquez 1982, Goldman et al. 1986 and Saleh 1987).

The lack of cross-resistance between the contact effect of OP's and pyrethroids and the larvicidal effect of *Bacillus thuringiensis* exotoxin in the house fly is encouraging. In 1983 field trials were made with Muscabac in Danish cattle, pig and hen farms, and in 1986 in pig farms with a wettable powder formulation of Muscabac (Jespersen and Skovmand 1984, Jespersen 1987). In some farms an immediate inhibitory effect on house fly larval development was seen, but a persistent efficacy was not observed. The persistence of the effect of Muscabac was at a maximum eight weeks if treated manure from pigs, poultry and calves were left undisturbed in a regulated room in a two month trial, however, if 5% or 10% of the manure was replaced weekly by fresh untreated manure to simulate natural conditions, the effect of Muscabac decreased rather soon (Jespersen 1986). A more persistent effect of Muscabac was observed in wet manure compared to dry manure (Jespersen and Skovmand, 1984).

From field trials in Finland with a *Bacillus thuringiensis* preparation containing spores, endotoxin and exotoxin Tulisalo and Rautapää (1983) reported a slower but longer-lasting effect on the house fly populations in cowhouses and piggeries; however, in some of these farms aerosol insecticides were also used. In a tropical environment Carlberg et al. (1985) found the flies *Chrysomyia* sp. and *Eristalis* sp. sensitive to treatment with Muscabac, and they obtained a good lasting control of flies in many latrines, but control was poor or partial in some, especially those in which fresh feces accumulated rapidly. If effective, *Bacillus thuringiensis* exotoxin for control of larvae of the house fly might be useful in view of the insecticide resistance problems in house fly control; however, until now a long lasting effective control has not been demonstrated in animal units. It might be possible to increase the efficacy of Muscabac by using higher dosages in field situations or to improve the conditions for growth of the *Bacillus thuringiensis* in the manure.

REFERENCES CITED

Carlberg, G., C. M. Kihamia and J. Minjas. 1985. Microbial control of flies in latrines in Dar es Salaam with a *Bacillus thuringiensis* (serotype 1) preparation, Muscabac. Mircen Journal 1: 33-44.

Carlberg, G. and R. Lindström. 1981. Possibility of fly resistance to *Bacillus thuringiensis* exotoxin (*Drosophila melanogaster*). British Crop Prot. Conf., Proceed. 1981: 591-597.

_____. 1987. Testing fly resistance to thuringiensin produced by *Bacillus thuringiensis*, serotype H-1. J. Invertebrate Pathol. 49: 194-197.

Finney, D. J. 1971. Probit analysis. Third edition. Cambridge University Press, London.

Georghiou, G. R. and M. G. Vasquez. 1982. Assessing the potential for development of resistance to *Bacillus thuringiensis var. israelensis* toxin (BTI) by mosquitoes. Mosquitoes Control Research Annual Report (USA) 1982: 80-81.

Goldman, I. F., J. Arnold and B. C. Carlton. 1986. Selection for resistance to *Bacillus thuringiensis* subspecies *israelensis* in field and laboratory populations of the mosquito *Aedes aegypti*. J. Invertebrate Pathol. 47: 317324.

Guillet, P. 1984. La lutte contre l'onchocercose humaine et les perspectives d'integration de la lutte biologique. Entomophaga 29 (2): 121-132.

Harvey, T. L. and D. E. Howell. 1965. Resistance of the house fly to *Bacillus thuringiensis* Berliner. J. Invertebrate Pathol. 7: 92-100.

Ignoffo, C. M. and R. T. Roush. 1986. Susceptibility of permethrin - and methomyl - resistant strains of *Heliothis virescens* (*Lepidoptera: Noctuidae*) to representative species of entomopathogens. J. Econ. Ent. 79: 334-337.

228

Iman, M., D. Soekarna, J. Situmorang, I. M. G. Adiputra and I. Manti. 1986. Effect of insecticides on various field strains of diamondback moth and its parasitoid in Indonesia. Diamondback Moth Management. Proc. First Internat. Workshop, Tainan, Taiwan 1985: 313-323.

Jespersen, J. B. 1986. Houseflies. Investigations on biological control agents. *Bacillus thuringiensis var. thuringiensis.* Danish Pest Inf. Lab. Ann. Rep. for 1985:

____. 1987. Houseflies. Investigation on biological control agents. *Bacillus thuringiensis var. thuringiensis.* Danish Pest Inf. Lab. Ann. Rep. for 1986: 51-52.

Jespersen, J. B. and 0. Skovmand. 1984. *Bacillus thuringiensis* laboratory and field evaluation for control of *Musca domestica.* Danish Pest Inf. Lab. Ann. Rep. for 1983: 3738.

Keiding, J. 1977. Resistance in the houseflies in Denmark and elsewhere. In D.L. Watson & W.A. Brown (edit.) 1977, Pesticide management and Insecticide Resistance. Acad. Press: 261-302.

Keiding, J. and J. B. Jespersen. 1986. Effect of different control strategies on the development of insecticide resistance by houseflies: Experience from Danish farms. 1986 British Crop Protection Conference - Pests and Diseases: 623-630.

McGaughey, W. H. 1985. Insect resistance to the biological insecticide *Bacillus thuringiensis.* Science 229: 193195.

McGaughey, W. H. and R. W. Beeman. 1988. Resistance to *Bacillus thuringiensis* in colonies of Indianmeal moth and almond moth (*Lepidoptera: Pyralidae*). J. Econ. Ent. 81 (1): 28-33.

McGaughey, W. H. and D. E. Johnson. 1987. Toxicity of different serotypes and toxins of *Bacillus thuringiensis* to resistant and susceptible Indianmeal moths (*Lepidoptera: Pyralidae*). J. Econ. Ent. 80 (6): 1122-1126.

Rupes, V., J. Ryba, H. Hanzlova, and J. Weiser. 1987. The efficiency of beta-exotoxin of *Bacillus thuringiensis* on susceptible and resistant house fly. In J. Olejnicek (edit.) Medical and Veterinary Dipterology, Proc. Internat. Conf. Ceske Budejovice 1987: 262-265.

Saleh, M. S. 1987. Effect of larval selection with two bioinsecticides on susceptibility levels and reproductive capacity of *Aedes aegypti* (L.). Anzeiger f. Schadl. kunde, Pflanzenschutz, Umweltschutz, 60: 55-57.

Stone, T. B., S. R. Sims and P. G. Marrone. 1989. Selection of tobacco budworm for resistance to a genetically engineered *Pseudomonas*-fluorecens containing the delta endotoxin of *Bacillus-thuringiensis-ssp-kurstaki*. J. Invertebr. Pathol. 53 (2): 228-234.

Sun, C. N., G. P. Georghiou and K. Weiss. 1980. Toxicity of *Bacillus thuringiensis var. israelensis* to mosquito larvae variously resistant to conventional insecticides. Mosq. News 40: 614-618.

Tulisalo, U. and J. Rautapää. 1983. The control of flies in cowhouses and swineries with *Bacillus thuringiensis* - preparation. Växtskyddsnotiser 47: 1-2, 15-22.

Wilson, B. H. and E. C. Burns. 1968. Induction of resistance to *Bacillus thuringiensis* in a laboratory strain of house flies. J. Econ. Ent. 61: 1747-1748.

Yang, Z. Q., J. F. Liu and W. D. Liu. 1985. Studies on the insecticide resistance of *Culex tritaeniorhynchus* Giles in China. Contrib. Shanghai Inst. Entomol. 5: 123-130.

Yeh, R., A. Whipp and J. P. Trijan. 1986. Diamondback moth resistance to synthetic pyrethroids: how to overcome the problem with deltamethrin. Diamondback Moth Management. Proc. First Internat. Workshop Tainan, Taiwan 1985: 379-386.

17. *Entomophthora muscae*
(Entomophthorales: Entomophthoraceae)
as a Pathogen of Filth Flies

Bradley A. Mullens

ABSTRACT

The fungal pathogen *Entomophthora muscae* (Cohn) Fresenius is known to infect a number of species of filth-dwelling Diptera, though we have few data to quantify levels of activity in these hosts. Literature is reviewed regarding the life cycle and taxonomic status of *E. muscae* s.l. Seasonal activity in *Musca domestica* L. on a southern California dairy was essentially the same as that previously reported for poultry operations in the region, with fall epizootics that reached infection levels of 60-80%. *Stomoxys calcitrans* L. adults were not infected by the pathogen in the field, in spite of being collected in an area and time of high pathogen activity in *M. domestica*. A laboratory study was done to assess the impact of *E. muscae* on fecundity in *M. domestica*. Flies which were exposed to the pathogen at <36 h of age did not oviposit before they died. Flies exposed at 48-96 h of age laid only 20% as many eggs as uninfected females from the same cohort over their lifespan, in spite of developing and holding mature eggs. Thus *E. muscae* appears to modify host behavior in *M. domestica*, as previously noted in *Delia* and *Psila*. Some areas of needed research are noted.

Introduction. Most research on natural enemies of filth-breeding muscoid flies has dealt with predators and parasites of the immature stages (see Axtell & Rutz 1986). There has been relatively little work on pathogens, either of immature or adult flies. Probably the most widely recognized pathogen of adult flies is the fungus *Entomophthora muscae* (Cohn) Fresenius. In spite of numerous accounts in the literature reporting this fungus from a variety of filth flies (see Greenberg 1971, Mullens et al. 1987), we have had no quantitative data regarding the impact of this pathogen in these hosts until recently. The prevalence of *Entomophthora muscae* on southern California poultry facilities, as well as a general discussion of the pathogen, already has been presented by Mullens et al. (1987), and the reader is referred to that publication for more detailed data and discussion. I would like to present previously unpublished data and update the pertinent literature here.

Life cycle. Some review of the life cycle of *E. muscae* is appropriate. Infected flies generally are killed after an incubation period of five to eight days. I have observed incubation periods as short as four days or as long as eleven to twelve days at temperatures of 20-22°C. Pathogen strain and host species, as well as host size, age, and level of pathogen exposure, all can affect incubation period (Mullens 1985, Steinkraus & Kramer 1987, Mullens 1989). Incubation period is dependent on temperature as well (Carruthers & Haynes 1985, Eilenberg 1987a). Critically ill flies cannot fly during the last few hours of life, and have a tendency to crawl upwards on vegetation or structures, sometimes until they encounter some small obstacle. Death usually occurs within a few hours of sunset, though time of death is influenced by photoperiod regime (Mullens, unpublished data).

Once the host has been killed, conidiophores of the fungus make their way through soft regions of the abdomen, particularly the intersegmental membranes and ventral surface. Occasionally conidiophores may emerge from other body areas, such as the occiput. In some pathogen strains, rhizoids, or "holdfasts" are produced from the end of the labellum, which is extended and attached to the substrate (Belazy 1984). Wings and legs typically also are outstretched, giving the cadaver a very distinctive pose. In *Musca domestica* L. at 20-22°C, discharge of conidia begins within a few hours of host death, reaches the highest level about ten to twelve hours after host death, and is completed generally within twenty hours of death (Mullens & Rodriguez 1985). Discharge may be increased by, but does not require, saturated humidities (Kramer 1980, Mullens & Rodriguez 1985, Carruthers & Haynes 1986), and duration and intensity of discharge are influenced by the temperature and host involved (Carruthers & Haynes 1986, Eilenberg 1987a).

Primary conidia carry a coating of cytoplasm and wall material and may travel a distance of several cm when discharged. They tend to adhere to objects near the cadaver. Primary conidia which fail to strike a host can form and discharge secondary conidia, which also are infective. Secondary conidia form long germ tubes under saturated conditions, such as likely occur in the micro-environment near the fly cuticle, and are probably the infective forms (Kramer 1980, Carruthers & Haynes 1986).

Host penetration probably is accomplished through a combination of enzymatic and physical means (Roberts & Aist 1984), but more work is needed on this critical aspect. Penetration can occur almost anywhere on the host, though it most frequently is through the abdomen. Hyphal bodies can be found in the haemocoel within 28 h of exposure and proliferate in the abdomen and fat body; most internal tissues are destroyed by the time death occurs (Brobyn & Wilding 1983).

Taxonomic status. The taxonomic status of *E. muscae* currently is unresolved. It certainly appears to be a species complex. Strains from various regions and hosts vary in conidial size, but also in the number and diameter of nuclei present in each conidium (Keller 1984). There has been some speculation as to which of the potential entities in the group is the pathogen which was described originally (Cohn 1855). Unfortunately, the number and size of nuclei were not noted in the original description, though size and shape of the conidia were. Based on size, Keller (1984) felt that his group A, which has an average of 4.4-6.6 nuclei per conidium, was perhaps the fungus of Cohn. Steinkraus and Kramer (1988) recently published an amplified description of *E. scatophagae* Giard, and noted that this form has twelve to nineteen nuclei (fifteen average), which agreed with Keller's preliminary assignment of this pathogen to group C. In Denmark, the pathogen found most commonly in *M. domestica* on pig farms has four to eight nuclei, though another form with an average of fourteen to fifteen nuclei also occurs occasionally in that host (U.S. Olesen, Danish Pest Infestation Laboratory, personal communication). The relative prevalence of the form with fewer nuclei in *M. domestica* in Denmark supports the hypothesis that the fungus with which Cohn worked in Germany may have been Keller's group A.

I have collected *E. muscae*-infected *Pollenia rudis* (F.) in southern California in May that had four to six nuclei per conidium, but have not found this form in *M. domestica* or other filth flies on poultry or dairy operations in California. In California, four manure-breeding muscoid fly hosts that share the same superficial habitat on poultry ranches have twelve to twenty (usually fourteen to eighteen) visible nuclei per conidium following staining with aceto-orcein. There is great variability in the infectivity of these 'host strains' from one host to another, however (Mullens 1989). Until such time as the various forms of *E. muscae* are characterized more fully, it is important that workers try at least to document some of the more obvious characteristics such as gross morphology (e.g. conidial size) and particularly the number and size of nuclei. It also is extremely desirable to deposit live voucher material in a place designed to handle and store it pending attempts to characterize the isolates. The best location of which I am aware is the USDA-ARS Insect Pathology Research Laboratory, Boyce Thompson Institute, Ithaca, NY.

Entomophthora muscae Activity on Poultry and Dairy Operations in California

We still do not have a great deal of data regarding activity of *E. muscae* in filth fly populations. In temperate regions in America and Europe the pathogen tends to occur in fall epizootics in *M. domestica* populations (e.g. Brefeld 1871, Graham-Smith 1916, Ystrom 1980, Mullens et al. 1987). Much less is known of its activity in other species of filth flies, but it may infect a variety of species in the laboratory (Steinkraus & Kramer 1987). Mullens et al. (1987) noted *E. muscae* epizootics in *M. domestica* (fall), *Fannia canicularis* (L.) (spring), and *Ophyra aenescens* (Weidemann) (winter, spring, fall) on southern California caged layer operations. Infection levels could reach 80% or more at some sites. The pathogen appeared to be density-dependent and was limited by high temperatures (above 25-30°C). Different strains may differ in their infectivity, operational temperature ranges, host range, etc., and much more work remains to be done in this area. There may be promise in releasing exotic strains with superior characteristics (e.g. temperature tolerance, infectivity) in a particular region for improved fly control, or possibly inoculating or augmenting endemic pathogens to improve control. *Entomophthora muscae* is much less limited by low humidity than are most other fungi in this group. Kramer & Steinkraus (1987) released *E. muscae*-infected and healthy *M. domestica* into a small (4m x 3m x 2.5m) sealed poultry building (no birds present) in New York and thus infected the healthy flies therein. While their suggestion that this demonstrates control potential clearly is premature, it does demonstrate that transmission occurred outdoors in a large cage-type setting.

During 1982 and 1983, fly populations were monitored over a period of two years at a dairy in the Chino Basin near Riverside, California. The sampling methodology is discussed fully in Mullens et al. (1987). Flies were collected with a sweep net every one to two weeks from hay bales, feed aprons, commodity pits, etc. The primary fly species were *M. domestica* and *Stomoxys calcitrans* L. Most of the *S. calcitrans* were collected in the spring, which agrees with their seasonal distribution on dairies in the area (Mullens & Meyer 1987). Over 100 were collected in the fall from hay bales which had large numbers of *M. domestica* cadavers killed by *E. muscae*. Of a total 687 *S. calcitrans* held which lived for the seven day holding period, none was observed with patent signs of *E. muscae* infection. Those which died during the holding period were not dissected, but none showed any signs of mycosis. This supports the premise that stable flies probably are not susceptible to *E. muscae*, though it is possible that a host-adapted strain exists. All the flies of which I am aware that are susceptible to *E. muscae* in nature have sponging-lapping mouthparts with large labellar lobes. These may be an integral part of the fungus-cadaver attachment mechanism. Steinkraus & Kramer (1987) did succeed in transmitting *E. muscae* from *M. domestica* to *S. calcitrans* in the laboratory, but only four of ninety exposed flies showed any indication of mycosis, and even those produced few conidia and had atypical postmortem signs.

The seasonal pattern of mycosis in *M. domestica* is presented in Figure 1. The fall epizootic pattern clearly was visible, with maximum infection levels reaching 60-80% (Figure 1a). This pattern basically was identical to that observed in *M. domestica* on caged layer operations in the region (Mullens et al. 1987), except that the level of infection seen on the dairy was higher than that seen on most of the poultry operations. The sample size for each collection is presented in Figure 1b. This is a rough measure of seasonal distribution of *M. domestica* activity; flies were relatively scarce during winter and spring and were abundant during summer and fall. The population density of *M. domestica* likely was more than sufficient for an *E. muscae* epizootic during the summer, but high temperatures limited pathogen activity, as was the case on the poultry operations (Mullens et al. 1987). We made no attempt to quantify cadaver distribution at this site, but cadavers commonly were seen during epizootic periods, particularly on support beams for structures and on stacks of hay bales.

A substantial amount of work has been done in the field on *E. muscae* in *M. domestica* populations on pig farms and other livestock settings in Denmark over the last eight to ten years. Abstracts of the work may be found in the yearly reports from the Danish Pest Infestation Laboratory (e.g. Ystrom 1980, Olesen 1985). The pathogen clearly is a substantial mortality factor in late summer and fall. Olesen, particularly, has done some very interesting work regarding high temperature limitations of *E. muscae* and interactions

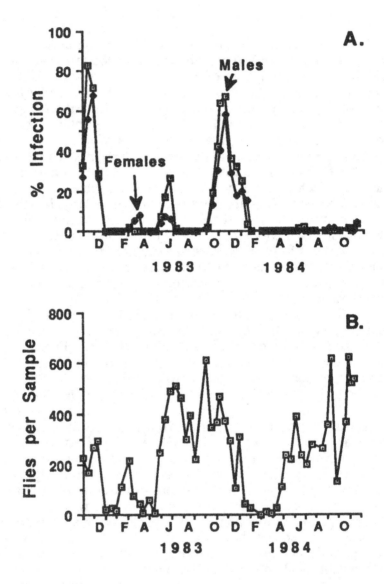

Figure 1. Seasonal prevalence of *Entomophthora muscae* in *Musca domestica* on a southern California dairy (A.) and number of flies assayed on each date (B.).

with fly behavior (Olesen, personal communication), which hope-fully will be published in greater detail soon.

Effects of *Entomophthora muscae* on Fly Reproduction

Naturally, the potential of *E. muscae* as a biological control agent rests in part on the role of the pathogen in reducing the repro-ductive success of infected hosts. Given an incubation period of five to eight days (substantially longer at cooler temperatures), infected *M. domestica* females conceivably could complete development of at least one batch of eggs. Larsen & Thomsen (1940) studied the minimum preoviposition period of *M. domestica* (provided with milk and bread as food and fresh horse manure for oviposition) at a range of temperatures. Representative times were approximately as follows (extracted from their Table 5): 35°C- 1.8 days, 30°C- 2.4 days, 25°C- 3.1 days, 22°C- 4.6 days, 19°C- 5.9 days, 15°C- 9.9 days, 13°C- 16.6 days. Average times are longer. I have held *E. muscae*-infected *M. domestica* at a range of temperatures to determine the incubation period, following an initial infection period of 24 h at 21°C. More carefully controlled experiments are pending, and it is known that minor differences in incubation period ($\pm$ 1-2 days) can be influenced by fly age, size (and thus sex), and level of pathogen exposure (Mullens 1985). The approximate incubation periods for a California strain of *E. muscae* were as follows (five replications of thirty infected flies each per temperature): 34°C- no conidia produced, 30°C- 5.4 days, 26°C- 5.0 days, 22°C- 6.2 days, 18°C- 6.3 days, 14°C- 9.1 days, 10°C- 17.6 days. It thus appears that the fungus development procedes more rapidly than oogenesis at low temperatures, while oogenesis procedes considerably more rapidly at higher temperatures (>22°C). In fact, the theoretical threshold for development is 12.2°C in *M. domestica* (Larsen & Thomsen 1940), while Carruthers and Haynes (1985) determined a theoretical deveopmental threshold temperature of 5°C for *E. muscae* from the onion fly, *Delia antiqua* (Meigen). Relative speed of oogenesis versus fungal development is only one aspect of the relationship, however. I am aware of two studies that have examined oviposition behavior of *E. muscae*-infected Diptera. Carruthers (1981) held field-collected *D. antiqua* in the laboratory and made daily observations of oviposition, as well as when death occurred due to mycosis. Infected and uninfected flies produced similar numbers of eggs for approximately the first three days, but the infected flies then ceased oviposition, in spite of the fact that death did not occur for another two to three days. He estimated a 50% reduction in fecundity for infected flies. Eilenberg (1987b) demonstrated that carrot flies, *Psila rosae* L., infected with *E. muscae* on days one to four of adult life were unable to lay their eggs near the food plant. Additionally, the number of eggs produced by infected females was reduced substantially, and a number of females did not oviposit at all. Eggs laid by infected females were fertile and developed normally when placed on suitable media.

I conducted an experiment to determine whether *E. muscae* might alter oviposition by infected *M. domestica*. Two groups of flies were used: 1) <36 h post-eclosion (young) and 2) two to five days post-eclosion (old). Flies had access to food and water from the time of eclosion. Half the females in each age group were infected with *E. muscae* at day zero according to Mullens (1986), while the others were not exposed. The four treatments thus were 1) young, healthy; 2) young, infected; 3) old, healthy; 4) old, infected. Flies were held at 21°C in 3.8 liter containers at a 14L:10D photoperiod and were supplied with dried milk, sugar, and water. Four containers (replications) were used per treatment. Each container contained fifty females and fifty healthy males, for a total of 200 females per treatment. Each day a cup with tissues soaked with milk (oviposition substrate) was placed in each container in the morning; it was removed at the end of the day (8h later) to determine the number of eggs volumetrically. Dead flies were counted and removed daily. Also, two live females were removed from each container per day (n=8/treatment/day) and dissected to assess the stage of follicle maturation (Trepte & Trepte-Feuerborn 1980). Monitoring egg production and mortality continued for ten days; the last dissections were done on day six.

The old flies, both healthy and infected, began ovipositing on day one. Oviposition for the two old fly groups was similar until day three, when the healthy flies began laying many more eggs (Figure 2). Most females in the old, infected group (94%) were dead by day seven (most died on day six and seven), and laid a total of only 3,060 eggs. The old, healthy females, on the other hand, laid a total of 16,380 eggs by day seven and 24,570 through day ten. The eggs laid by the infected and healthy flies were viable, with hatching above 85%. The young, healthy flies began laying eggs on day three; they had laid 9,090 eggs by day six and 23,220 through day ten. In contrast, the young, infected flies laid no eggs at all (Figure 3), and 100% of the females were dead by day six (most died on day five and six).

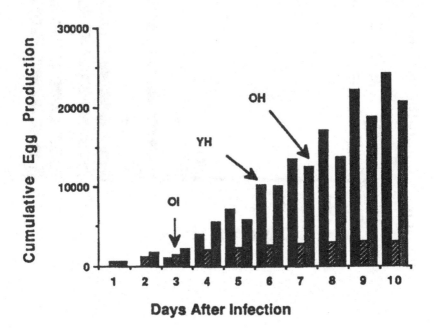

Figure 2. Cumulative egg production in *Musca domestica* for control flies and flies exposed to *Entomophthora muscae*. Fly groups as follows: YH = young (36 h old), healthy; YI = young, infected; OH = old (48-96 h old), healthy; OI = old, infected. n = 200 females per group.

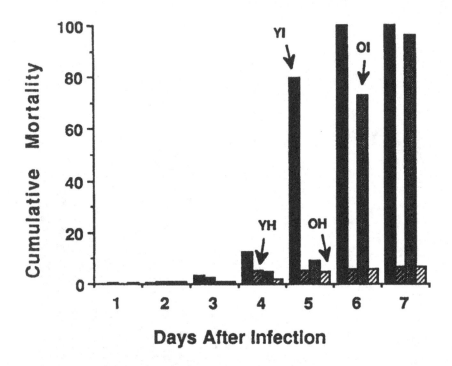

Figure 3. Cumulative mortality of *Musca domestica* exposed to *Entomophthora muscae* versus healthy flies from the same cohorts. See Figure 2 for abbreviations.

Ovarian dissections revealed that the primary follicles of the young flies on day one, on average, were at stage one, while those of the old flies were between stage three and four (Figure 4). Most females in both the old fly groups had fully developed follicles (stages five and six) by day three. The young, healthy flies developed eggs by day four and five (though a very few already had oviposited on day three). The young, infected females, on the other hand, often did not completely mature their eggs (Figure 4). By day four and five, it appeared that the follicles were being resorbed in many of the females, and the haemocoel at this point was packed with hyphal bodies of the fungus.

From these studies it appears that infection with *E. muscae* may alter oviposition behavior in *M. domestica*, just as already has been demonstrated for *Delia* and *Psila* (loc. cit.). The old, infected females developed and held apparently full complements of eggs. Over the first four days of the experiment, when mortality in both old, healthy and old, infected flies was equal and <5%, egg deposition by infected females was only 21% (2,070 eggs) of the total laid by healthy females (9,540 eggs). The effect was more marked in young, infected females, some of which developed eggs but none of which laid them. I have not yet examined what effect, if any, *E. muscae* may have on males or mating behavior or success.

Concluding Remarks

Entomophthora muscae is very much in need of further evaluation. When I first began working on the fungus six years ago, I was quite surprised to see how often it had been observed and how little it had been studied. Its seasonal occurrence is relatively easy to document with consistent monitoring (Mullens et al. 1987), and more of this should be done. It quite likely is an important natural mortality factor in many areas and in a number of filth-breeding muscoid fly species. Because it likely is a complex, we definitely are in need of someone with the proper mycological background and biochemical expertise to characterize the forms. Also, because the pathogen is relatively easy to maintain (with vigilance and practice) in vivo in the laboratory, many avenues of study on pathogen-host interactions are open to investigation.

242

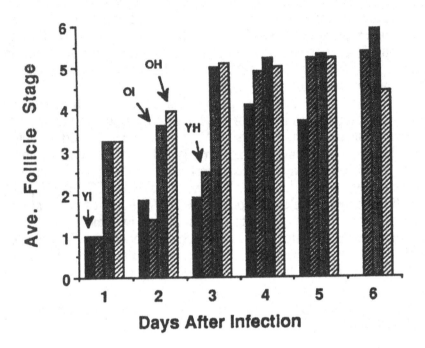

Figure 4. Average stage of primary follicle development in *Musca domestica* exposed to *Entomophthora muscae* versus development in healthy flies from the same cohorts. See Figure 2 for abbreviations.

REFERENCES CITED

Axtell, R. C. & D. A. Rutz. 1986. Role of parasites and predators as biological fly control agents in poultry production facilities. pp. 88-100. In Patterson, R. S. & D. A. Rutz (eds.). Biological control of muscoid flies. Misc. Publ. Entomol. Soc. Amer. 61. 174 pp.

Belazy, S. 1984. On rhizoids of *Entomophthora muscae* (Cohn) Fresenius (Entomophthorales: Entomophthoraceae). Mycotaxon 19: 397-407.

Brefeld, O. 1871. Untersuchungen uber die Entwicklung der *Empusa muscae* und *Empusa radicans*, und die durch sie verursachten Epidimien der Stubenfliegen und Raupen. Abhandl. d. naturf. Ges. Halle. 12: 1-52.

Brobyn, P. J. & N. Wilding. 1983. Invasive and developmental processes of *Entomophthora muscae* infecting houseflies (*Musca domestica*). Trans. Br. Mycol. Soc. 80: 1-8.

Carruthers, R. I. 1981. The biology and ecology of *Entomophthora muscae* (Cohn) in the onion agroecosystem. Ph.D. Dissertation, Michigan St. Univ., East Lansing.

Carruthers, R. I. & D. L. Haynes. 1985. Laboratory transmission and in vivo incubation of *Entomophthora muscae* (Entomophthorales: Entomophthoraceae) in the onion fly, *Delia antiqua* (Diptera: Anthomyiidae). J. Invert. Pathol. 45: 282-287.

_____. 1986. Temperature, moisture, and habitat effects on *Entomophthora muscae* (Entomophthorales: Entomophthoraceae) conidial germination and survival in the onion agroecosystem. Environ. Entomol. 15: 1154-1160.

Cohn, F. 1855. *Empusa muscae* und die Krankheit der Stubenfliegen. Verhandl. Kaiserl. Leop. Carol. Akad. Naturforsch. 25: 301-360.

Eilenberg, J. 1987a. The culture of *Entomophthora muscae* (C) Fres. in carrot flies (*Psila rosae* F.) and the effect of temperature on the pathology of the fungus. Entomophaga 32: 425-435.

_____. 1987b. Abnormal egg-laying behavior of female carrot flies (*Psila rosae*) induced by the fungus *Entomophthora muscae*. Entomol. Exp. Appl. 43: 61-65.

Graham-Smith, G. S. 1916. Observations on the habits and parasites of common flies. Parasitol. 8: 440-544.

Greenburg, B. 1971. Flies and disease. Vol. 1. Princeton Univ. Press, N. J. 856 pp.

244

Keller, S. 1984. *Entomophthora muscae* als Artencomplex. Mitt. Schweiz. Entomol. Ges. 57: 131-132.

Kramer, J. P. 1980. The house fly mycosis caused by *Entomophthora muscae*; influence of relative humidity on infectivity and conidial germination. J. N. Y. Entomol. Soc. 88: 236-240.

Kramer, J. P. & D. C. Steinkraus. 1981. Culture of *Entomophthora muscae* in vivo and its infectivity for six species of muscoid flies. Mycopathologia 76: 139-143.

_____. 1987. Experimental induction of the mycosis caused by *Entomophthora muscae* in a population of house flies (*Musca domestica*) within a poultry building. J. N. Y. Entomol. Soc. 95: 114-117.

Larsen, E. B. & M. Thomsen. 1940. The influence of temperature on the development of some species of Diptera. Vidensk. Medd. fra Dansk naturh. Foren. 104: 1-75.

Mullens, B. A. 1985. Host age, sex and pathogen exposure level as factors in the susceptibility of *Musca domestica* to *Entomophthora muscae*. Entomol. Exp. Appl. 37: 33-39.

_____. 1986. A method for infecting large numbers of *Musca domestica* (Diptera: Muscidae) with *Entomophthora muscae* (Entomophthorales: Entomophthoraceae). J. Med. Entomol. 23: 457-458.

_____. 1989. Cross-transmission of *Entomophthora muscae* (Zygomycetes: Entomophthoraceae) among naturally infected muscoid fly (Diptera: Muscidae) hosts. J. Invert. Pathol. 53: 272-275.

Mullens, B. A. & J. A. Meyer. 1987. Seasonal abundance of stable flies (Diptera: Muscidae) on California dairies. J. Econ. Entomol. 80: 1039-1043.

Mullens, B. A. & J. L. Rodriguez. 1985. Dynamics of *Entomophthora muscae* (Entomophthorales: Entomophthoraceae) conidial discharge from *Musca domestica* (Diptera: Muscidae) cadavers. Environ. Entomol. 14: 317-322.

Mullens, B. A., J. L. Rodriguez & J. A. Meyer. 1987. An epizootiological study of *Entomophthora muscae* in muscoid fly populations on southern California poultry facilities, with emphasis on *Musca domestica*. Hilgardia 55: 41 pp.

Olesen, U. S. 1985. *Entomophthora muscae* infecting houseflies. Danish Pest Infest. Lab. Ann. Rep. 1985 (86): 56-57.

Roberts, D. W. & J. R. Aist (eds.). 1984. Infection processes of fungi.The Rockefeller Foundation, A Bellagio Conference, March 21-25, 1983. 196 pp.

Steinkraus, D. C. & J. P. Kramer. 1987. Susceptibility of sixteen species of Diptera to the fungal pathogen *Entomophthora muscae* (Zygomycetes: Entomophthoraceae). Mycopathologia 100: 55-63.

_____. 1988. *Entomophthora scatophagae* desc. ampl. (Zygomycetes: Entomophthorales), a fungal pathogen of the yellow dung fly, *Scatophaga stercoraria* (Diptera: Anthomyiidae). Mycotaxon 32: 105-113.

Trepte, H. H. & C. Trepte-Feuerborn. 1980. Development and physiology of follicular atresia during ovarian growth of the house fly, *Musca domestica*. J. Insect Physiol. 26: 329-338.

Ystrom, P. 1980. Biology of houseflies: mortality factors. Danish Pest Infest. Lab. Ann. Rep. 1980 (81): 60.

18. Susceptibility of Muscoid Fly Parasitoids to Insecticides Used in Dairy Facilities

Donald A. Rutz and Jeffrey G. Scott

Present house fly *Musca domestica* L. and stable fly *Stomoxys calcitrans* (L) control techniques rely heavily on the use of insecticides, and these insecticides plus application-related costs constitute a significant production expense for dairy producers (Lazarus et al. 1989). In addition to cost, the traditional reliance on chemical control as a unilateral pest suppression tactic has other serious drawbacks including the development of insecticide resistance by the target pests and the destruction of nontarget biological control agents. These problems have generated considerable interest in the development of integrated fly management programs for dairy facilities with emphasis on biological control agents.

Two years ago we conducted a survey of wild house fly populations associated with dairy production throughout New York to determine the prevalence and severity of pest fly resistance to insecticides that are typically used for fly control on dairy farms (Scott et al. 1989). All ten wild populations that were sampled statewide showed high levels of resistance to all of the insecticides currently used for fly control, with the exception of permethrin and Vapona. In many cases these assays substantiated farmers' reports of treatment failures. As economic realities within the agrichemical industry indicate that few new materials can be expected for fly suppression in coming years, it is more imperative than ever that the existing chemicals be used judiciously to slow the spread of resistance and minimize the hazards to natural enemies of the target pest species.

Research during the past several years on the development of the biological component of integrated fly management programs for dairy facilities has shown that indigenous hymenopterous parasitoids (Family Pteromalidae) of fly pupae are important biological control agents of muscoid flies and play an essential role in pest population regulation at these facilities (Petersen and Meyer 1983; Smith and Rutz 1985, Miller and Rutz 1990). Unfortunately, little information is available on the comparative susceptibility of house fly and stable fly biological control agents to insecticides that are commonly used in dairy facilities for fly control. Clearly, the use of insecticides that both aggravate existing fly resistance problems and are highly toxic to the fly parasitoids is cost-ineffective in the short (inadequate pest control) as well as the long run (increasingly high doses needed and eventual "loss" of the insecticide to resistance). In contrast, selective use of insecticides that are efficacious against flies and relatively benign to the parasitoids would be cost effective and prolong the life of the most effective insecticides. The preservation of effective integrated pest management programs for flies, therefore, requires an understanding of the interactions among pesticide, target pests, and associated beneficial insects.

In this chapter, we summarize the published insecticide susceptibility test results of *Spalangia cameroni* Perkins and *Urolepis rufipes* (Ashmead) and present new data for two other parasitoids, *Muscidifurax raptor* Girault and Sanders and *Pachycrepoideus vindemmiae* (Rondani).

MATERIALS AND METHODS

The parental stocks of the four parasitoids species used in the study were collected from dairy farms in and around Cayuga, Schuyler and Tompkins counties of New York State. Parasitoids were collected by using the sentinel pupae parasitoid monitoring method (Rutz and Axtell 1979) and colonized in the laboratory for ten to twelve months prior to testing.

Seven technical grade insecticides were tested: crotoxyphos (Ciodrin®, Fermenta Animal Health), dichlorvos (Vapona®, Fermenta Animal Health), dimethoate (Cygon®, American Cyanamid), fenvalerate (DuPont), permethrin (Cooper Animal Health), Pyrenone®, (pyrethrins + piperonyl butoxide, Fairfield American) and tetrachlorvinphos (Rabon®, Fermenta Animal Health). The synergists piperonyl butoxide (PBO >95%), diethyl maleate (DEM, Aldrich Chem. Co., 97%) and S,S,S-tributyl phosphorotrithioate (DEF) were also used.

Parasitoids (1-7 day old) were tested by a surface contact assay. A glass vial (37 cm^2 surface area) was treated with insecticide in acetone (0.5 ml), and the acetone was evaporated while rotating the vial to provide uniform coverage of all the glass surfaces. The parasitoids to be tested were aspirated and transferred into the treated vial. The vial was covered with a wire mesh top and the parasitoids were given a small drop of 10% sucrose *(S. cameroni* and *P. vindemmiae)* or honey *(M. raptor* and *U. rufipes)* on the lid as a food source. Males and females were tested separately. Susceptibility was assessed after holding the parasitoids at a constant temperature (±28°C) for 48 h with a photoperiod of 15:9 (L:D). For each species twenty parasitoids were tested at each dose with at least four doses, giving between 0 and 100% mortality. The dosage of the synergists was 100 ug/jar for DEM and 10 ug/jar for PBO and DEF. These doses gave no control mortality. Each test was replicated a minimum of four times. Results were pooled, and Abbott's formula (Finney 1971) was used to correct for control mortality. Data were analyzed by a computer program based on the method of Finney (1971), as adapted for PC use by Raymond (1985).

To investigate the basis of the differential susceptibility of males compared to females, we determined the average weight of each sex for each of the parasitoid species. Groups of twenty male or female parasitoids of each species (n = 30) were weighed immediately after they were killed with cyanide gas. Statistical analysis was by Student's t test. Furthermore, for *U. rufipes* we determined the amount of radiolabeled permethrin taken up by each sex. Uptake and distribution of ^{14}C-permethrin were determined by placing twenty male or twenty female *U. rufipes* in a glass vial (the same as those used for the bioassays) treated with 10,000 dpm ^{14}C-permethrin. No mortality was observed at this dose. Specific testing procedures are presented in Scott and Rutz (1988).

RESULTS

We (Scott et al. 1988) determined that the toxicity of the seven insecticides to *S. cameroni* decreased in the order of Cygon > Vapona ≥ Ciodrin > Pyrenone ≃ permethrin > fenvalerate (Table 1). Overall, there was a 126-fold reduction in toxicity from Cygon to fenvalerate based on female LC$_{50}$ values. Although there was substantial variation in the toxicity of compounds within classes of

Table 1. Toxicity of Seven Insecticides to Male and Female S. *cameroni*.[a]

Insecticide	Sex	LC$_{50}$[b]	95% Confidence Intervals		Slope (S.E.)	n
			Lower Limit	Upper Limit		
Rabon	male	c	–	–	–	1580
	female	c	–	–	–	2400
Vapona	male	3.1	2.9	3.4	2.8 (0.2)	900
	female	1.5	1.4	1.6	2.8 (0.2)	1380
Ciodrin	male	1.1	1.0	1.2	5.6 (0.1)	680
	female	1.8	1.7	1.9	8.3 (0.6)	560
Cygon	male	0.12	0.11	0.13	5.0 (0.5)	500
	female	0.14	0.13	0.15	6.5 (0.5)	600
Permethrin	male	1.5	1.4	1.7	2.5 (0.2)	1040
	female	5.5	5.2	5.8	5.2 (0.4)	680
Pyrenone	male	5.0	4.7	5.4	4.2 (0.3)	700
	female	5.0	4.7	5.3	5.4 (0.6)	440
Fenvalerate	male	11.3	10.2	12.4	2.8 (0.2)	700
	female	17.6	16.4	18.9	3.5 (0.3)	820

a From Scott et al. (1988)
b In units of ug/vial
c Data did not fit a straight line

insecticides, the pyrethroids were generally less toxic than the organophosphates to *S. cameroni*. For all insecticides except Rabon, the data points fit the log dose - probit line (P ≤ 0.05, Chi-squared test; P ≥ 0.10 for Vapona) and there was no indication of any plateaus. A plot of log dose Rabon vs. mortality (probit scale) shows the existence of a plateau that extends over a 16-fold dose range (Figure 1) for both sexes. This indicates the existence of two distinct populations, having different susceptibilities to Rabon. The most likely explanation for this phenomenon was that 45% of the population was resistant to Rabon due to selection in the field. The similar response of males and females suggested that the resistance was not affected by the chromosomal ploidy in these parasitoids.

In an effort to determine the mechanism(s) responsible for *S. cameroni* resistance to Rabon, we examined the effect of three synergists: PBO, DEF and DEM, inhibitors of oxidative, hydrolytic and glutathione S-transferase mediated metabolism, respectively (Scott et al. 1988). In these tests the effect of the synergist was evaluated against both the susceptible (i.e., the ca. 55% of the individuals that die at the lower doses) and resistant (i.e., the ca. 45% of the individuals that die only at the higher doses) populations (Figure 2). DEM did not affect either population. DEF showed a noticeable but erratic effect. DEF had no effect on the susceptible population, and its effect on the resistant population was inconclusive. PBO had a significant effect on both populations. There was an 8-fold increase in toxicity to the susceptible population, suggesting that monooxygenases are important in the detoxification of Rabon in *S. cameroni*. There was also a significant effect on the resistant population (i.e., greater than seen for the susceptible strain), implying that the resistance may be due in part to elevated monooxygenase activity.

With *U. rufipes* we (Scott and Rutz 1988) observed that the toxicity of the insecticides decreased in the order of Cygon > permethrin > Rabon > fenvalerate > Pyrenone ≃ Vapona > Ciodrin with a 46-fold reduction in toxicity between Cygon and Ciodrin based on female LC_{50} values (Table 2). Toxicity of compounds within classes of insecticides varied substantially, suggesting no obvious generalization about toxicity based on class of insecticide. For all insecticides the probit line was relatively steep (slope > 2), and the data points fit the probit line (P ≤ 0.05, chi-squared test) with no indication of any plateaus. Thus, this strain of *U. rufipes* appears to be genetically homogeneous.

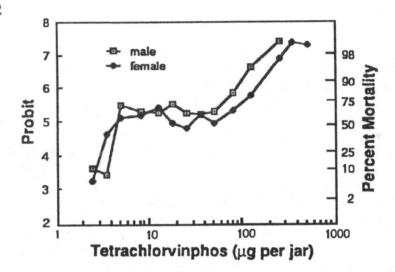

Figure 1. Toxicity of tetrachlorvinphos to male and female *S. cameroni.*

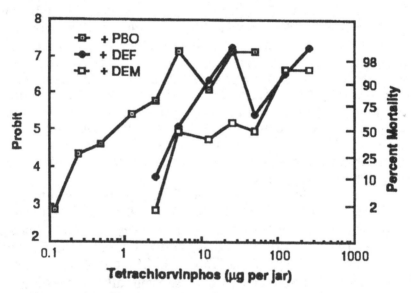

Figure 2. Toxicity of tetrachlorvinphos to female *S. cameroni* in the presence of the synergists piperonyl butoxide (PBO), DEF or diethyl maleate (DEM).

Table 2.   Toxicity of Seven Insecticides to Male and Female _U. rufipes._[a]

Insecticide	Sex	LC$_{50}$[b]	95% Confidence Intervals		Slope (S.E.)	n
			Lower Limit	Upper Limit		
Rabon	male	0.39	0.37	0.42	4.7 (0.6)	460
	female	1.0	1.0	1.0	5.3 (0.5)	520
Vapona	male	1.3	1.2	1.4	3.7 (0.3)	780
	female	1.7	1.5	1.9	1.9 (0.3)	820
Ciodrin	male	1.1	1.0	1.2	4.0 (0.4)	680
	female	3.0	2.9	3.2	6.9 (0.7)	620
Cygon	male	0.02	0.02	0.03	4.7 (0.5)	520
	female	0.07	0.06	0.07	6.1 (0.7)	380
Permethrin	male	0.19	0.17	0.21	3.0 (0.4)	560
	female	0.34	0.32	0.36	4.4 (0.4)	640
Pyrenone	male	1.3	1.2	1.4	6.5 (0.7)	420
	female	1.6	1.5	1.7	4.6 (0.4)	920
Fenvalerate	male	0.88	0.78	0.98	2.7 (0.4)	520
	female	1.4	1.2	1.5	2.8 (0.3)	1060

[a] From Scott and Rutz (1988)
[b] In units of ug/vial

The toxicity of the seven insecticides (based on female LC$_{50}$ values) to *P. vindemmiae* decreased in the order of Cygon > Ciodrin > Rabon $\geq$ permethrin $\simeq$ Pyrenone > Vapona > fenvalerate (Table 3). A 179-fold reduction in toxicity from Cygon to fenvalerate was observed. With the exception of Vapona, the pyrethroids were generally less toxic to *P. vindemmiae* than the organophosphates. Similar to *U. rufipes*, this strain of *P. vindemmiae* also appears to be homogeneous based on the facts that the probit line was relatively steep (slope > 2) and the data points for all insecticides fit the probit line (P $\leq$ 0.05, chi-squared test).

The toxicity of the insecticides to *M. raptor* varied dramatically (277-fold) from the most toxic (Cygon) to the least toxic (fenvalerate, Table 4). Specifically, the toxicity of the insecticides to *M. raptor* decreased in the order of Cygon > permethrin > Ciodrin > Pyrenone > Rabon > Vapona $\simeq$ fenvalerate. Similar to *U. rufipes*, the toxicity of the compounds within classes of insecticides varied substantially, suggesting no obvious toxicity generalization based on class of insecticide. This strain of *M. raptor* also appears to be homogeneous.

In all cases, males of *U. rufipes* and *M. raptor* were significantly (P $\leq$ 0.05) more susceptible than females by an average of ca. 2-fold (range 1.2 to 2.9) (Table 5). *S. cameroni* and *P. vindemmiae* females were generally more tolerant than males except in the cases of Pyrenone and Vapona where there were no significant differences between the sexes. In addition, no significant differences between the sexes in susceptibility to Cygon were observed in *S. cameroni*. The general trend of males being more sensitive to insecticides compared with females correlates well with differences in size between the sexes. For all parasitoid species, males were found to be consistently smaller (P $\leq$ 0.05) than females, with average male and female weights of 0.64 $\pm$ 0.02 mg and 0.74 $\pm$ 0.04 respectively for *S. cameroni*, 0.78 $\pm$ 0.2 mg and 1.3 $\pm$ 0.03 mg respectively for *U. rufipes*, 0.27 $\pm$ 0.01 mg and 0.39 $\pm$ 0.01 mg respectively for *P. vindemmiae* and 0.36 $\pm$ 0.01 mg and 0.56 $\pm$ 0.02 mg respectively for *M. raptor*. No significant difference was observed between male and female *U. rufipes* in the uptake of ^{14}C-permethrin (Scott and Rutz 1988). Therefore, the general greater sensitivity of the males appears to be due to their picking up a higher dose per mg body weight than females. The lack of consistent difference in insecticide susceptibility between male and female *S. cameroni* is probably due to the body weights of

Table 3. Toxicity of Seven Insecticides to Male and Female *P. vindemmiae.*

Insecticide	Sex	LC$_{50}$[a]	95% Confidence Intervals		Slope (S.E.)	n
			Lower Limit	Upper Limit		
Rabon	male	2.3	2.2	2.4	10.6 (1.2)	400
	female	2.8	2.7	3.0	7.8 (0.7)	500
Vapona	male	13.5	10.5	21.6	2.1 (0.6)	200
	female	10.4	9.0	11.5	5.0 (0.9)	180
Ciodrin	male	1.2	1.1	1.3	4.5 (0.3)	680
	female	1.7	1.6	1.8	3.7 (0.4)	660
Cygon	male	0.08	0.07	0.09	5.0 (0.5)	520
	female	0.11	0.10	0.12	5.9 (0.5)	640
Permethrin	male	2.0	1.8	2.0	2.2 (0.1)	560
	female	3.4	3.2	3.7	4.0 (0.3)	1040
Pyrenone	male	3.4	3.1	3.9	3.3 (0.4)	640
	female	3.8	3.5	4.1	4.3 (0.3)	840
Fenvalerate	male	15.2	14.2	16.3	4.7 (0.3)	820
	female	19.7	18.5	20.9	4.8 (0.3)	880

a In units of ug/vial

Table 4. Toxicity of Seven Insecticides to Male and Female *M. raptor*.

| Insecticide | Sex | LC$_{50}$[a] | 95% Confidence Intervals | | Slope (S.E.) | n |
			Lower Limit	Upper Limit		
Rabon	male	2.6	2.5	2.8	5.7 (0.5)	620
	female	3.3	3.1	3.5	5.0 (0.4)	700
Vapona	male	6.9	6.2	7.7	2.4 (0.2)	940
	female	14.6	13.3	15.9	2.6 (0.2)	920
Ciodrin	male	0.37	0.34	0.42	2.6 (0.2)	920
	female	1.1	1.0	1.2	4.6 (0.4)	680
Cygon	male	0.04	0.04	0.04	5.3 (0.5)	520
	female	0.06	0.05	0.06	6.8 (0.5)	740
Permethrin	male	0.10	0.09	0.11	3.2 (0.3)	500
	female	0.21	0.19	0.23	3.7 (0.3)	820
Pyrenone	male	1.3	1.2	1.3	5.8 (0.5)	760
	female	2.0	1.9	2.2	4.0 (0.3)	720
Fenvalerate	male	10.4	9.8	11.1	3.8 (0.2)	1120
	female	16.6	15.4	18.1	3.5 (0.3)	780

a In units of ug/vial

Table 5. Effect of Parasitoid Sex on Susceptibility to Seven Insecticides Used on Dairies.

Insecticide	Female : Male LC$_{50}$ ratio				
	S. cameroni	U. rufipes	P. vindemmiae	M. raptor	
Cygon	1.2[a]	2.6	1.4	1.5	
Ciodrin	1.6	2.8	1.4	2.9	
Permethrin	3.7	1.8	1.7	2.1	
Pyrenone	1.0[a]	1.3	1.1[a]	1.6	
Rabon	–[b]	2.5	1.2	1.2	
Vapona	0.5[a]	1.3	0.8[a]	2.1	
Fenvalerate	1.6	1.5	1.3	1.6	

[a] No significant difference observed between females and males
[b] Data did not fit a straight line

males and females being more similar in this species compared to the other species.

DISCUSSION

In summary, all parasitoid species were found to be extremely susceptible to Cygon and, with the exception of *U. rufipes*, least susceptible to fenvalerate (Figure 3). The pyrethroids were generally less toxic than the organophosphates to *S. cameroni*. The toxicity of compounds within classes of insecticides varied substantially in *U. rufipes, P. vindemmiae* and *M. raptor* and, therefore, no obvious generalization can be made about toxicity based on class of insecticide in these species. Overall, *U. rufipes* was the most insecticide sensitive of the four parasitoid species tested.

A possible explanation for the higher susceptibility of *U. rufipes* to insecticides may be their microhabitat association. In a study of the microhabitat associations of parasitoids attacking house fly pupae at dairy farms in central New York State (Smith 1989), *U. rufipes* was observed to be most abundant in outdoor fly breeding areas. Whereas, *M. raptor* was common in both indoor and outdoor fly breeding areas and *S. cameroni* and *P. vindemmiae* were most abundant indoors. Unfortunately, because *U. rufipes* was the only outdoor species tested, it is unclear as to whether this overall high sensitivity to insecticides was due to having limited exposure to insecticides or other factors.

Recent years have witnessed the development of a substantial data base on nontarget effects of pesticide applications in crop agroecosystems (e.g., Pickett 1988, Theiling and Croft 1988). In contrast, prior to our studies, nearly no data were available for livestock production systems. Bartlett (1963) investigated the contact toxicity of pesticide residues to five species of hymenopterous parasitoids, one of which was a house fly parasitoid, *Spalangia drosophilae*, and six species of coccinellid predators. He reported that *S. drosophilae* was susceptible to over 80% of the sixty-one pesticides tested. There has been only one other very limited investigation on the susceptibility of muscoid fly pupal parasitoids to insecticides used for fly control (Hahn 1979). In this study two fly pupal parasitoids, *M. raptor* and *S. nigroaenea*, were tested. The parasitoids were most susceptible to malathion followed by permethrin, Cygon, and Rabon, and least susceptible to Vapona.

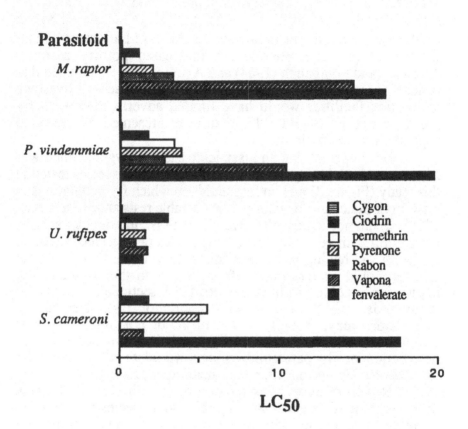

Figure 3. Comparative susceptibility of parasitoids to insecticides used on dairies.

In field situations, Axtell (1970a) reported that insecticide applications over the entire manure surface in poultry facilities resulted in destruction of fly natural enemy populations and only short-lived (seven to ten days) fly control so that frequent repeated applications became necessary. In more recent studies, Axtell and Edwards (1983) in North Carolina, and Meyer et al. (1984) in California, observed little or no adverse effects of Larvadex® (Ciba-Geigy) on predator populations in treated poultry manure. Anderson and Poorbaugh (1964) and Axtell (1970b) concluded that residual insecticide applications to fly resting surfaces in livestock production facilities would have limited adverse effects on fly predators and parasitoids. No data were presented, however, to support these conclusions.

It is apparent that an insecticide such as Cygon, which was found to be both highly toxic to all four parasitoid species tested in this study (Figure 3) and an insecticide to which fly populations in New York State have developed considerable resistance (Scott et al. 1989), would not be cost effective. In contrast, insecticides such as permethrin with generally lower frequencies of resistance (Scott et al. 1989) and having less deleterious effects on certain parasitoid species would be more cost effective due to the combined fly population regulation effect of both the insecticides and the fly parasitoids. Continued farmer reports of fly resistance to insecticides may, in fact, be the result of not only the actual development of resistance by the fly populations, but also extensive destruction of fly parasitoids which would result in rapid fly population build-up soon after insecticide application.

Results of our studies demonstrate that there are substantial dose-response differences among parasitoid species to the active ingredients in insecticides that are commonly used to manage muscoid flies. Additional research, being conducted under simulated field conditions, is in progress using label-rate premise treatments of formulated products to determine whether the inherent toxicological differences among parasitoid species are large enough to be manifested under field conditions.

With information available on house fly insecticide resistance (Scott et al. 1989), parasitoid microhabitat preferences (Smith 1989), parasitoid seasonal activity patterns (Miller and Rutz 1990, Smith 1989) and insecticide susceptibility, we will be able to tailor our fly management recommendations according to the time of year and the primary fly breeding areas on the farm. As fly populations warrant, we will be able to recommend a particular insecticide to which flies express the least amount of resistance and that is least destructive to particular parasitoid species which may be prevalent on a farm due to either their seasonality or microhabitat preferences.

Obviously we have not completely solved the fly management problems on dairy farms in New York with this study. However, we are adding an extremely important component to our integrated fly management programs. Information of this type is also critical for the development of accurate predictive fly management models.

The evolving system of new dairy production and management practices and new insecticide chemistry will continue to challenge our ability to effectively manage flies on livestock farms. Continued urbanization of our agricultural areas makes this challenge even more critical.

ACKNOWLEDGMENTS

The authors wish to thank J. Walcott, B. Schanbacher, and C. Stokes for technical assistance. This article is a publication of the Cornell University Agricultural Experiment Station, New York State College of Agriculture and Life Sciences. This research was supported by Hatch projects 139414 and 139428 and a grant from the Northeast Pesticide Impact Assessment Project.

LITERATURE CITED

Anderson, J. R. and J.H. Poorbaugh. 1964. Observations on the ethology and ecology of various Diptera associated with northern California poultry ranches. J. Med. Entomol. 1:131-147.

Axtell, R. C. 1970a. Fly control in caged-poultry houses: Comparison of larviciding and integrated control programs. J. Econ. Entomol. 63:1734-1737.

_____. 1970b. Integrated fly control program for caged-poultry houses. J. Econ. Entomol. 63:400-405.

Axtell, R. C. and T.D. Edwards. 1983. Efficacy and nontarget effects of Larvadex as a feed additive for controlling house flies in caged-layer poultry manure. Poultry Sci. 62:3472-2377.

262

Bartlett, B. R. 1963. The contact toxicity of some pesticide residues to hymenopterous parasites and coccinellid predators. J. Econ. Entomol. 56:694-698.

Finney, D. J. 1971. Probit Analysis. Third edition. Cambridge University Press, Cambridge. 318 pp.

Hahn, L. C. 1979. The comparative susceptibility of the house fly and two hymenopterous pupal parasites to five insecticides used in poultry houses. M.S. Thesis, North Carolina State University, Raleigh. 40 pp.

Lazarus, W. F., D. A. Rutz, R. W. Miller and D. A. Brown. 1989. Costs of existing and recommended manure management practices for house fly and stable fly (Diptera:Muscidae) control on dairy farms. J. Econ. Entomol. 82:1145-1151.

Meyer, J. A., W. F. Rooney and B. A. Mullens. 1984. Effect of Larvadex feed-through on cool-season development of filth flies and beneficial Coleoptera in poultry manure in southern California. Southwestern Entomol. 9:52-55.

Miller, R. W. and D. A. Rutz.. 1990. Survey of house fly pupal parasitoids on dairy farms in Maryland and New York. IN Biocontrol of Arthropods Affecting Livestock and Poultry. D.A. Rutz and R.W. Patterson, eds. Westview Press, Boulder, CO.

Petersen, J. J. and J. A. Meyer. 1983. Host preference and seasonal distribution of pteromalid parasites (Hymenoptera: Pteromalidae) of stable flies and house flies (Diptera: Muscidae) associated with confined livestock in eastern Nebraska. Environ. Entomol. 12:567-571.

Pickett, J. A. 1988. Integrating use of beneficial organisms with chemical crop protection. Phil. Trans. Royal Soc. London B 318:203-211.

Raymond, M. 1985. Presentation d'une programme Basic d'analyse log-probit pour micro-ordinateur. Cah ORSTOM ser Ent. med. et Parasitol. 23:117-121.

Rutz, D. A. and R. C. Axtell. 1979. Sustained releases of *Muscidifurax raptor* (Hymenoptera: Pteromalidae) for house fly (*Musca domestica*) control in two types of caged layer poultry houses. Environ. Entomol. 8:1105-1110.

Scott, J. G. and D. A. Rutz. 1988. Comparative toxicities of seven insecticides to house flies (Diptera: Muscidae) and *Urolepsis rufipes* (Ashmead) (Hymenoptera: Pteromalidae). J. Econ. Entomol. 81:804-807.

Scott, J. G., D. A. Rutz and J. Walcott. 1988. Comparative toxicity of seven insecticides to adult *Spalangia cameroni* Perkins. J. Agric. Entomol. 5:139-145.

Scott, J. G., R. T. Roush, and D. A. Rutz. 1989. Insecticide resistance of house flies (Diptera: Muscidae) from New York dairies. J. Agric. Entomol. 6:53-64.

Smith, L. 1989. The influence of microhabitat on the distribution and attack rate of hymenopteran parasitoids of house fly pupae at dairies in New York State. Ph.D. thesis. Cornell Univ. 187 pp.

Smith, L. and D. A. Rutz. 1985. The occurrence and biology of *Urolepsis rufipes*(Hymenoptera: Pteromalidae), a parasitoid of house flies in New York dairies. Environ. Entomol. 14:365-369.

Theiling, K. M. and B. A. Croft. 1988. Pesticide side-effects on arthropod natural enemies: a database summary. Agriculture, Ecosystems and Environment 21:191-218.

19. Computer Simulation Modeling of Fly Management

R. C. Axtell and R. E. Stinner

Management of arthropod pests of livestock and poultry is a complex process requiring a mixture of control strategies according to the species of pests and the different types of animal production systems (Axtell 1981, 1986a). Computer simulation models are needed to effectively deal with these complexities. Integrated pest management (IPM) is a part of the total livestock management system; the design and implementation of management techniques requires analysis of the pest population dynamics, among other factors. Critical to practical IPM is pest population monitoring and the use of pest population predictive models together with management predictive models to elucidate the appropriate management actions (Figure 1).

Biocontrol is a major management option which is becoming more important as the application of IPM principles increases. Greater use of biocontrol is also encouraged by the trend to low-input sustainable agriculture (LISA) along with desires to minimize the use of chemicals in animal production systems. Population dynamics models of biocontrol agents are as important as models of the pests themselves in the development of livestock and poultry IPM programs.

Computer simulation models of pest management in livestock and poultry production are needed to organize and visualize the complexities and interactions which are too numerous and subtle to be comprehended without assistance. Further, models allow predictions of the consequences of environmental and management changes. The validity of a simulation model, however, depends on the accurate conceptualization of the system, the reliability of the input data, and the proper mathematical expressions for components of the model and for manipulation of data. It is inevitable that lack of data and knowledge on relationships requires

265

266

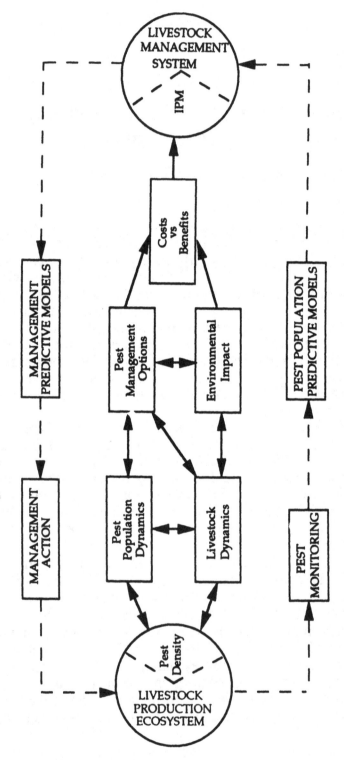

Figure 1. Diagram of the major components in the development and operation of an IPM program for livestock (including poultry) production systems. Adapted from Axtell (1981, 1986a).

the use of assumptions in the modeling effort. Such assumptions must be reasonable and explicit.

Filth fly management, among the pest problems of livestock and poultry production, is amenable to computer simulation modeling (Weidhaas 1986). The need is substantial due to the complex mixture of biological, cultural, and chemical methods required for fly management. Further, fly management is a problem common to all confined-animal production systems throughout the world. These systems present manure-management problems and an ideal habitat for fly production. Because confined-animal production systems are artificial ecosystems, designed and managed by humans, they are especially adaptable to the application of the IPM concept and to the use of pest management modeling. The increase in confined systems for dairy, swine, and poultry has intensified the filth fly problem, with the house fly, *Musca domestica*, being the most abundant pest species. Therefore, we have chosen to conceptualize and model house fly management. Our approach is based on the situation in caged poultry production, but with modifications is applicable to fly management in other types of confined-animal production systems as well as to other fly species.

MODELING CONCEPT

In our overall concept of house fly management, as shown in Figure 2, the population of adult flies in a confined-animal facility is variable due to the following major factors: 1) immigration/ emigration, 2) habitat conditions (temperature and moisture) in the house and in the accumulated manure, 3) the fly management practices used, including cultural (mainly timing and methods of manure handling and disposal), biological (enhancement of the population of natural enemies), and chemical (type and application methods), and 4) the populations of natural enemies in the habitat. The natural enemies or "biocontrol agents" include predators, parasites, pathogens, and competitors. In poultry houses, predators (*Macrocheles* mites and *Carcinops* beetles) and pteromalid parasites (*Muscidifurax* and *Spalangia*) are especially important. There are likely to be considerable interactions among the species of natural enemies and among the other components in the fly management conceptualization. Likewise, the impacts of management strategies are interrelated. Further, the habitat conditions in the housing results from modifications of the ambient temperature and moisture

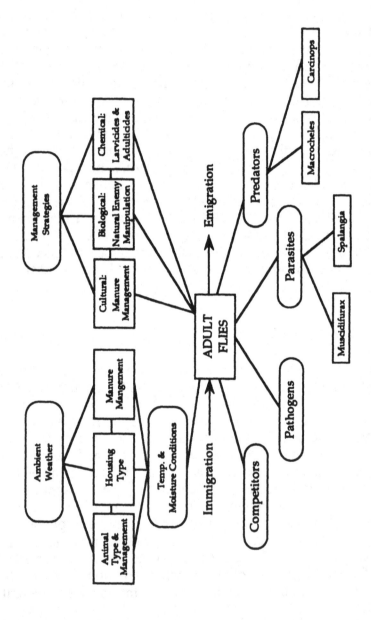

Figure 2. Major components in a model of house fly management for confined-animal production systems.

by the interacting impacts of animal type and management, type of housing, and system of manure handling and disposal. Figure 2 conveys the major relationships, without inclusion of all possible interactions and interrelationships, in order to provide a practical framework for fly management modeling.

Biological Components. Understanding the biology and behavior of the house fly and its natural enemies is basic to developing a realistic fly management model. Identification and quantification of the sources of fly mortality (in all life stages) are critical to the elucidation of the fly population dynamics. Further details and references on the ecology and management of the house fly and its natural enemies are provided by Axtell 1986b,c; Axtell and Arends 1990; Patterson and Rutz 1986; Rueda and Axtell 1985. Briefly, some of the major considerations are as follows.

The house fly life cycle (Figure 3) in poultry manure requires six to ten days in a warm house. The eggs are deposited in batches in manure that has the most attractive odor and moisture level. Eggs hatch within a day to first-instar larvae (L1), which undergo development and molting through two additional instars (L2, L3). Larval stages move through the manure, though usually not very deeply owing to the anaerobic conditions. When there is no alternative, the larvae can tolerate rather liquid conditions. The third-instar larva develops into the pupal stage within its thickened integument (puparium). Pupation occurs in the drier portions of the manure near the surface and edges. An adult fly emerges from the puparium and disperses after hardening of the cuticle. It often remains in the vicinity if the habitat is conducive to feeding, mating, and oviposition. Adult flies spend considerable time on the surface of the manure in the daytime, but at night rest on surfaces, mostly in the upper parts of the poultry house.

Other species of flies (as well as certain beetles) may compete with development of the house fly in the manure (Axtell and Arends 1990). The soldier fly, *Hermetia illucens,* has robust larvae which churn the manure and physically render the habitat less suitable for the house fly. Larvae of *Ophyra* spp. flies occupy the same manure microhabitats as the house fly and also may prey on the house fly larvae. Under certain conditions, *Ophyra* flies may become the dominant species in poultry houses. It is unclear how much this may be due to competition versus predation on the house fly. Several species of *Fannia* larvae develop in the manure, occasionally become quite abundant, and conceivably may be competitors. In most circumstances, however, there appears to be

270

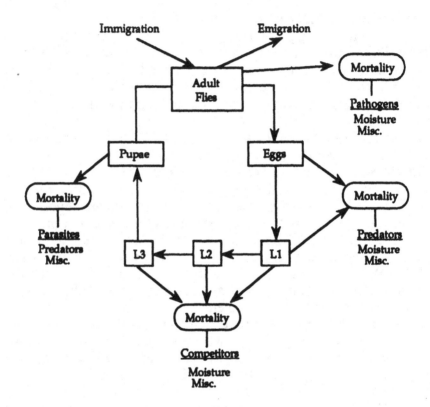

Figure 3. Diagram of house fly, *Musca domestica*, life cycle and factors affecting population size.

little evidence of house fly development being seriously impeded by competitors in the poultry manure.

Predators and parasites of the house fly are major suppressors of fly populations (Axtell and Rutz 1986). The most abundant predators on eggs and first-instar larvae of the house fly (and other muscids) are the mite, *Macrocheles muscaedomesticae*, and the histerid beetle, *Carcinops pumilio* (Geden and Axtell 1988, Geden et al. 1988). Mite and beetle predators are most abundant and effective in manure that is not too wet and accumulates in a distinct pile. These predators are most abundant at or near the crest and slightly beneath the surface of the pile, in the areas most likely to have house fly eggs and first-instar larvae. Mites develop rapidly (Figure 4), with only two to three days required under warm favorable conditions to complete the life-cycle from egg to adult (with intermediate larva, protonymph, and deutonymph stages). Most predation by mites is by the adult feeding on house fly eggs and first-instar larvae, with a preference for the eggs. Alternate foods for the adult and earlier life stages are nematodes, acarid mites, and other dipterous eggs and small larvae. Adult mites exhibit phoresy, with the house fly the most common carrier. This phoresy is regulated by olfactory responses to the flies and the aging manure, resulting in dispersal to the most favorable fly breeding areas.

The predator *C. pumilio* has a longer life cycle (Figure 5), requiring about forty days for development from egg to adult (with two intermediate larval stages and a pupal stage) in the warm conditions of a poultry house. Adults have a mean longevity of almost 100 days at 27°C. Adults (and to a much lesser extent, larvae) feed on eggs and first-instar larvae of the house fly. Alternate foods are eggs and small larvae of other Diptera in the manure. Beetles require an undisturbed site in the manure for normal pupation.

Hymenopterous parasites (Pteromalidae) are a major factor in destroying house fly pupae in poultry facilities (as well as in other livestock operations). Several species in the genera *Muscidifurax* and *Spalangia* are common (Rueda and Axtell 1985). Intergeneric differences in searching abilities and competitiveness exist. All species have the same basic life cycle (Figure 6) with development from egg to adult on the fly pupae (within the puparium) requiring about three weeks under poultry house conditions. One adult parasite emerges from each puparium and destroys the fly pupa during development. Adult parasites destroy additional flies by host feeding. Development on the fly pupae is through three larval

272

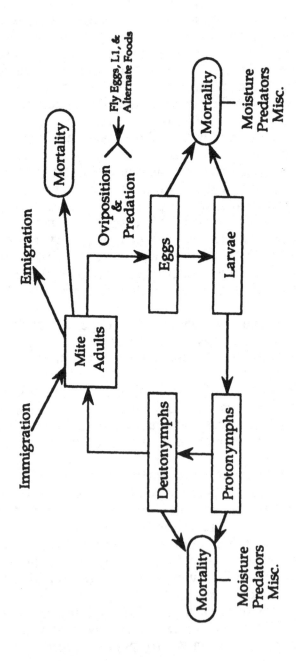

Figure 4. Diagram of macrochelid mite predator, *Macrocheles muscaedomesticae*, life cycle and factors affecting population size.

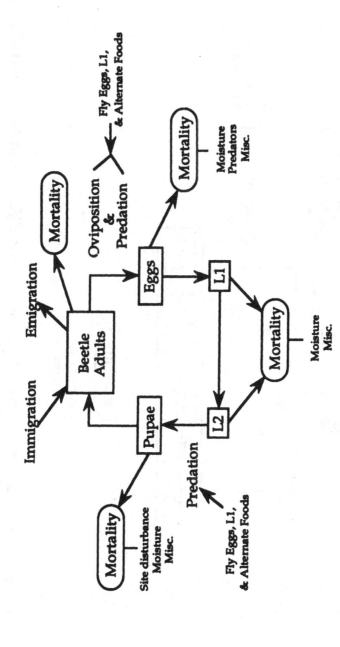

Figure 5. Diagram of histerid beetle predator, *Carcinops pumilio*, life cycle and factors affecting population size.

274

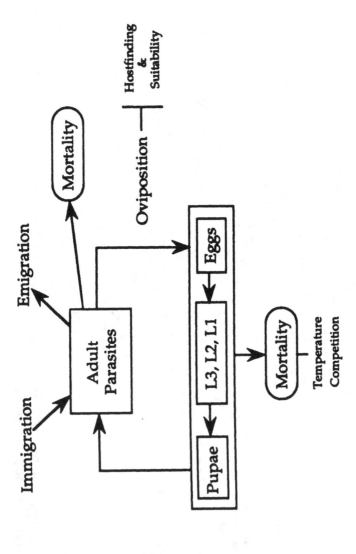

Figure 6. Diagram of pteromalid parasite (*Muscidifurax* and *Spalangia*) life cycle and factors affecting population size.

instars and a pupal stage. There are species of fly parasites in other genera that differ in their biology, particularly *Nasonia*, which develops several parasites on one fly pupa and *Tachinaephagus* (Encyrtidae), which parasitizes fly larvae. These genera of parasites are generally considered of less importance in suppressing house fly populations.

Fly pathogens have received less attention as suppressors of house fly populations in poultry houses. The fungus *Entomophthora muscae* is a common pathogen of adult flies (Mullens et al. 1987). With more research, it may become obvious that the population dynamics of it, and perhaps other pathogens, should be modeled. Since pathogens undoubtedly have some effect on fly mortality, they need to be included in the conceptualization of a fly management model.

Physical Components. Populations of the house fly and its natural enemies are temperature-dependent (Ables and Shepard 1976, Ables et al. 1976, Lysyk and Axtell 1987, Weidhaas et al. 1977). Although the temperature range in a poultry house is restricted to maintain suitable poultry management conditions, there is still considerable variation, which affects the arthropod population dynamics. Air temperature obviously directly affects the adult fly and parasite activities. Manure temperature is more directly related to immature stages of the house fly and parasites, and to all stages of the predaceous mites and beetles. Because it is more practical to routinely monitor air temperature, it is important to relate manure temperatures to air temperatures.

Moisture is also a major factor affecting populations of the house fly and its natural enemies (Fatchurochim et al. 1989). The manure moisture level influences the degree of oviposition by house flies and the survival of immature stages of the fly. It affects the survival of all stages of the predaceous mites and beetles. Manure that is very wet is unsuitable for survival of mites and beetles, while very dry manure does not support immature stages of the house fly. Air moisture (relative humidity) affects the rate of manure drying, and the activity and survival of adult flies and parasites. Relating moisture, both manure and air, to arthropod population dynamics is extremely difficult and only approximations and crude estimates are possible at this time.

Management Components. Management approaches to house fly (and other filth fly) control in poultry production are empirically understood. A judicious meshing of cultural, biological, and chemical techniques, along with appropriate monitoring of the fly populations are required (Axtell 1986a, Lysyk and Axtell

1986a,b). Although details vary with the type of housing, manure handling system, and climatic situation, the principles of fly management are universal. Cultural methods are primary and basically involve keeping the manure dry in order to discourage fly development and to encourage a heterogeneous population of predators and parasites that occur naturally and provide biological control of the fly population. Chemical control measures are supplements to cultural and biological methods. Since most insecticides that are effective against the house fly are also quite toxic to the predators and parasites, routine use of larvicides in the poultry manure is not compatible with maintenance of the biological control agents. Routine adulticides may be detrimental to the populations of adult parasites. Some selectivity is obtained by limited spot treatment of manure, residual surface treatments of areas where adult flies rest in the upper parts of the structure, and fly toxicant-baits. Timing of these control measures is facilitated by monitoring adult fly populations with standardized techniques, such as spot cards and baited jug-traps.

    <u>Submodels.</u> The overall conceptualization of house fly management requires relating the above considerations and others, in a scheme with several interacting submodels. Thus we include submodels for: the house fly (FLYMOD), mite predators (MACMOD), beetle predators (CARMOD), and pteromalid parasites (PARMOD), based on the conceptualizations presented in Figures 3, 4, 5, and 6, respectively.

    These population dynamics models are temperature-dependent and the relevant temperature will be that of the manure and/or air as described previously. Several alternatives for relating manure to air temperature are being developed by Wilhoit (personal comm.). Linkage of these submodels provides a base for input of the estimated impacts of various management strategies such as the timing and methods of manure removal, pesticide applications, and parasite and/or predator augmentation. Techniques for modeling this fly management scheme are available based on modeling efforts with other pest management problems, but selection of the most appropriate techniques and adapting them to this particular situation are critical.

MODELING METHODS

In modeling the population dynamics of the house fly, its predators and parasites, we have used a modification of the approach described by Wilkerson et al. (1986). This same paradigm is used for each of the species involved, so that many of the same subroutines are called by the different species' population generator ("bookkeeper"). This generator is a cohort "bookkeeper," using two arrays for each stage. The elements of the first array contain the densities per unit area, D, of individuals which entered that stage 0, 1, 2, 3, .. N time units ago (Figure. 7). N is the maximum number of time units an individual can remain in a given stage. Summing the elements gives the total density of individuals presently in the given stage. Within a stage, at the beginning of a new time step, individuals in a given element are multiplied by survival during that time and moved to the next ("older") element.

$$D(j,t+1) = D(j,t) * S(j, \Delta t) \qquad \text{Eq. (1)}$$

where: $D(j,t) =$ density of individuals in stage j, which entered that stage t time units ago.

$S(j, \Delta t) =$ survival of individuals in stage j during time Δt.

The second array contains elements, A, representing the median physiological age of the corresponding D element. Within a stage, at the beginning of a new time step, the median development rate, R, occurring during the time step is calculated, added to the previous physiological age, and the sum moved to the next element.

$$A(j,t+1) = A(j,t) + (R * \Delta t) \qquad \text{Eq. (2)}$$

where: $A(j,t) =$ median physiological age of individuals in stage j who are t time units old chronologically.

$\Delta t =$ time step.

The developmental increment, Q, which equals $R * \Delta t$, (and, thus, physiological age) is temperature dependent. We use median (50%) development rate rather than the mean rate because of the way in which developmental variability is modeled (see below). For virtually all poikilotherms, development rate is a nonlinear function

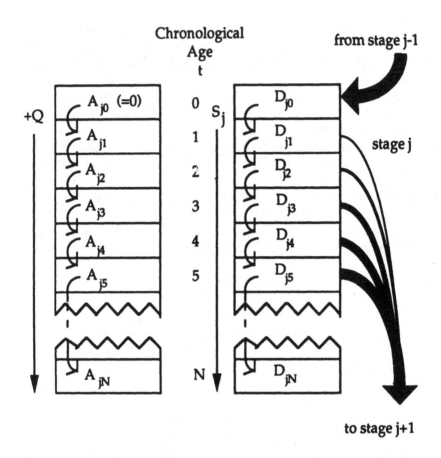

Figure 7. Density (D) and physiological age (A) compartmental flow used in population model. Q is developmental increment and S is survival. See text for explanation.

of temperature (Sharpe and DeMichele 1977, Wagner et al. 1984), although many other models have assumed the relationship to be linear (i.e., day-degree models). As an example, Figure 8 shows the median development rate, R, versus temperature, T, for *Spalangia endius* (Mann et al. 1990b), fit with the Sharpe-DeMichele (1977) model.

$$R = \frac{b1 * \dfrac{T}{298.15} * \exp\left[\dfrac{b2}{1.987}\left(\dfrac{1}{298.15} - \dfrac{1}{T}\right)\right]}{1 + \exp\left[\dfrac{b3}{1.987}\left(\dfrac{1}{b4} - \dfrac{1}{T}\right)\right]} \qquad \text{Eq. (3)}$$

where: R = median rate of development.

T = temperature (degrees Kelvin).

b1,b2,b3,b4 = estimated parameters.

Since this submodel is computationally expensive, in our population models we simply have an array of the rates (previously calculated) for temperatures from 1 to 45°C.

Median physiological age at time t, A_t, is simply the sum of the ratio over time. For a cohort beginning development at time = 0.

$$A_t = \sum_{i=0}^{t} R * \Delta t \qquad \text{Eq. (4)}$$

where: A_t = median physiological age at time t.

R = rate of development for temperature at time i, from Eq. (3)

Δt = time step

Eq. (4) is the equivalent of Eq. (2), taken over multiple time steps. It is important to recognize that for each median rate, there is an entire distribution of development times (Figure 9) for any temperature. If these distributions are plotted against median physiological age from Eq. 4, then all of these distributions collapse to a single distribution (Figure 10), dependent on median

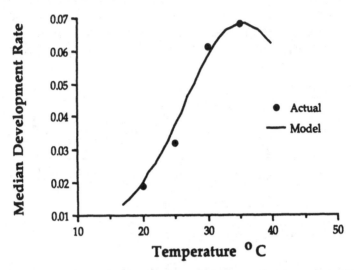

Figure 8. Median development rate as a function of temperature for *Spalangia endius*.

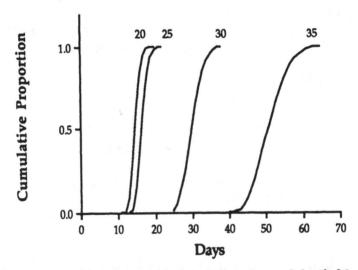

Figure 9. Cumulative proportion of individuals developed to adult for *Spalangia endius* for different temperatures.

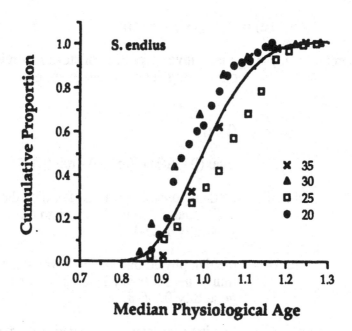

Figure 10. Cumulative proportion of individuals developed to adult for *Spalangia endius* as a function of physiological age, estimated from Eqs. 3 and 4.

physiological time. This "same-shape" assumption has been tested and is valid for all of the species involved in our efforts. Note that at a median physiological age of 1.0, 50% of the individuals have developed to the next stage. Given this, it is possible to calculate the proportion of individuals completing development, G, as a function of physiological age during time Δt, using the algorithm of Stinner et al. (1975):

$$G(j, \Delta t) = (P(j,t+1) - P(j,t)) \qquad \text{Eq. (5)}$$

where: $P(j,t) = $ cumulative proportion of stage j individuals completing development by time t.

$$P(t) = (1 - z)^{kz^2}$$

$z = (max(j) - A(j,t))/(max(j) - min(j))$

$max(j) = $ median physiological age at which the slowest individual in stage j develops to the next stage, generally 1.25-1.35.

$min(j) = $ median physiological age at which the fastest individual in stage j develops to the next stage, generally 0.7-0.8.

$k = $ empirical constant, calculated from max(j) and min(j) (see Stinner et al. 1975).

For immature *S. endius* development in our example, max(1), min(1), and k are 1.3, 0.8, and 2.1013, respectively.

In the models, transfer (development) to the next stage is based on the above, with all individuals entering the O element of the next stage density array, with their physiological age reset to zero. To account for individuals already molted, G must be divided by the proportion of individuals left (1 -P(j,t)):

$$D(j+1,O) = \sum_{t=1}^{N} D(j,t) * [G(j, \Delta t)/(1 - P(j,t))] \quad \text{Eq. (6)}$$

For example, with *S. endius* (Figure 8), the median rate of development, Rd, at constant 30°C is 0.06135. At 30°C, individuals (in different proportions) from 14 to 22 days old will molt to adults (Figure 9). Note that once 100% (G = 1.0) is reached, all the individuals have molted (and in the model transferred to the next stage). For chronological ages greater than 22 days, G is equal to one, but the corresponding densities of immatures older than 22 days are all zero (they have already molted). For 25°C, the median development rate is 0.03175 (Figure 8) and individuals from twenty-six- to thirty-eight-days-old molt, in different proportions, into adults (Figure 9). Since the development rate itself is lower, it takes longer to develop and 100% molting does not occur until day 38.

Obviously, for the adult to egg transfer:

$$D(1,O) = \sum_{t=1}^{N} D(\text{adult}, t) * S * F * G(\text{adult}, \Delta t) \quad \text{Eq. (7)}$$

where: $S =$ proportion females.

$F =$ lifetime fecundity.

$G(\text{adult}, \Delta t) =$ proportion of total eggs laid during Δt.

For any population dynamics model, development, immature survival, adult longevity, reproduction, and immigration/emigration must be considered. For predators and parasites, additional information on attack and parasitism processes are needed. Of these processes, only development, as discussed above, can be modeled as a general process for most species. How we have approached modeling these other processes for each species or species group is discussed below. Since both experiments and modeling efforts are still in progress, the exact functional forms of many of the relationships are subject to change. Many of the specific relationships are complex, based on data from multiple sources, and require detailed justifications and analyses beyond the scope of this chapter.

FLYMOD. In addition to temperature-dependent development, the house fly dynamics model includes: immature survival, adult longevity, and fecundity. To calculate survival rates, eggs and first instars are subject to predation by *Macrocheles muscaedomesticae* and *Carcinops pumilio* (Geden and Axtell 1988, Geden et al. 1988), and pupae are parasitized by pteromalid parasites of the genera *Spalangia* and *Muscidifurax* (Mann et al. 1990a,b). Additionally, all stages are assumed to be killed by temperatures below O°C. Immature survival is limited to between 35 to 75% moisture in the manure and outside this range, survival is assumed to be zero (Fatchurochim et al. 1989).

Adult flies can live a considerable time beyond their reproductive life. Thus, it is necessary to model both longevity and oviposition (Fletcher et al., submitted). Longevity is modeled as described for development. Median longevity rate, L, is defined as:

$$L = \exp(bl(T-b2)-1) \qquad \text{Eq. (8)}$$

where: T =   temperature (°C)

   bl, b2 = estimated parameters

Adult physiological age is calculated from Eq. (4), and adults die of "old age," using Eq. (5), with minimum age, maximum age, and k equal to 0.118, 2.496, and 1.8382, respectively. Reproduction (eggs/female/4 hours) is modeled as a function of current temperatures and physiological age. Since we are concerned with the population, and not individual females, it was assumed that inclusion of the oogenesis cycles of individuals is not necessary. Total fecundity (based on current 4-hour mean temperature) is calculated first. This relationship is described by Eq. (3), with total fecundity, F, replacing R. This value, multiplied by the proportion of that total fecundity which should be laid during the time interval, (Eq. (5)), based on the physiological age of each female adult cohort, simulates eggs laid/female during Δt. This result is then multiplied by the female adult density in each cohort and summarized over cohorts to provide total eggs laid during the 4-hour interval.

Ages and densities of eggs and larvae are recalculated every four hours, including production of new eggs and large larval transfer to the pupal stage. Because pupae and adults have considerably longer transit times (times in those stages), their physiological ages and survivals are updated only daily.

MACMOD. The mite *Macrocheles muscaedomesticae*, as a predator of eggs and first instars of house fly, has been the subject of considerable research which has facilitated modeling (see data and references in Geden and Axtell 1988, Geden et al. 1988, and Geden et al. 1990). A detailed description of the model, its assumptions, and data sources is presented in Geden et al. (1990), although a number of the functions used have already been revised (Wilhoit, pers. comm.). Immature survival is considered a function of crowding, habitat size, and habitat quality. Since it has not been feasible to determine when mortality occurs in the immature stages, survival proportion is simply accounted for as the immatures transfer to adults.

As with the house fly, this mite lives beyond its reproductive period. Adult aging and longevity are considered temperature-dependent only. The adult physiological longevity rate is calculated as proportional to a mean longevity of 21.2 days at 27°C. Mortality due to age is calculated from Eq. (5), with minimum and maximum ages set at 0.8 and 1.3, respectively.

Reproduction is calculated similar to the house fly, except that the resultant eggs/4-hour interval is reduced as a function of the quantity and quality of prey to allow for the effect of alternate prey. Additionally, mite crowding was found to decrease fecundity significantly, and a negative exponential, scaled 0-1, modifies the potential to actual reproduction. Potential lifelong fecundity, F, is defined as:

$$F = b1 * Z^{b2} * (1-Z)^{b3} \qquad \text{Eq. (9)}$$

where: $Z = (40-t)/(40-15)$

$t =$ temperature within the range of 40°C maximum and 15°C minimum

The proportion of eggs laid during a given time interval, P, is calculated using Eq. (5) (P = G of Eq. (5)), with minimum and maximum physiological ages set at 0.089 and 0.833, respectively. The maximum fecundity during a time interval, FEC, is thus:

$$FEC = F * P$$

This maximum fecundity is reduced by the effect of prey density and potential lower nutritional value of alternate prey, if available.

Details and justification for the above are presented in Geden et al. (1990).

Predation on house fly eggs and first instars is based on a maximum rate related to prey density at 27°C, with proportional changes determined by temperature (relative to 27°C), mite crowding, and deflection to alternate prey (if present).

As with the house fly, adult mites live considerably longer than the immatures. Therefore, immature development, reproduction and attack rates use a 4-hour time step, while adult aging and longevity have a 1-day time step.

CARMOD. Modeling of the histerid beetle, *Carcinops pumilio,* is in progress using an approach similar to that used for MACMOD (Wilhoit, pers. comm.). The major difference is in dealing with longevity, since the median adult longevity of the beetle is almost 100 days at 27°C. Because our prediction horizon is 30 days or less, adult aging is not considered in this model. Thus, fecundity, after the prereproductive phase, is considered constant, except for effects of temperature and beetle density.

Immature development, as related to temperature, is being described by Fletcher et al. (pers. comm.), using Eq. (3). Attack rates were determined by Geden and Axtell (1988) and Geden et al. (1988), with additional functional forms being explored by Wilhoit (pers. comm.).

PARMOD. The four pteromalid pupal parasites, *Spalangia endius, S. cameroni, Muscidifurax raptor,* and *M. zaraptor* have been the subject of considerable research (Rueda and Axtell 1985). Mann et al. (1990a,b) provide most of the data on temperature and density effects on development, parasitism, and pupal attack rates used in the parasite models. Surprisingly, all four species exhibited similar responses. Development is modeled in the same manner described for the other submodels. Immature parasite survival is assumed equal to fly pupal survival. Because the adult parasites host feed, in addition to ovipositing in fly puparia, separate values for pupal kill and parasite reproduction have to be calculated. A daily time step is used for both adults and immatures, but attacks occur with a 4-hour time step. A full description of the parasite model is being prepared.

For all four species, both pupal kill and parasitism are assumed to be functions of temperature and host/parasite ratios. In a series of experiments with both host and parasite densities varying (Mann et al. 1990a), no effect could be found of either host or parasite density alone that was not accounted for by using an asymptotic function of host/parasite ratio. That is, there were no apparent crowding effects. For all four species, both number killed/parasite, P_k, and number parasitized/parasite, P_p, are described by:

$$P_k \text{ or } P_p = \text{w1} \{1 - \exp[-\text{w2}(H/P)]\}) \quad \text{Eq. (10)}$$

where: w1 = temperature-dependent maximum asymptote calculated from Eq. (3);
w1 = R of Eq. (3)

w2 = estimated parameter

H/P = host pupa/adult parasite ratio

For all species, w1 for the P_k equation is greater than w1 for the P_p equation.

At this time, longevity is considered a function of temperature only, but it will be necessary to incorporate the findings of Ables and Sheppard (1976) that the longevity is affected by the pattern of reproduction during the adult life time. Since both immatures and adults are long-lived, compared to the immatures of other species in this overall model, a one-day time step is used for both.

An additional problem which arises with these parasites is that host pupae are susceptible to parasitism only when young (approximately the first 25% of their time as pupae). Mann's (1990a,b) experiments on density effects were conducted with young susceptible pupae only. Should older, nonsusceptible pupae be considered in density relationships (e.g., do parasites "waste" time probing these pupae)? Will parasites host-feed on the older pupae? Because the parasite responses are quite sensitive to density, these questions must be resolved.

Species interactions. Among the most difficult modeling problems to resolve is how to handle species interactions. For competing predators or parasites, during a time step, which species is allowed to attack first, or can rules be developed to partition fly hosts or prey appropriately within a time step? Is the present 4-hour time step sufficiently small that the order of attack makes no difference? Given the difficulty in executing multispecies validation studies, the answers to these questions are not easily obtained. Wilhoit (pers. comm.) is exploring model sensitivities to alternative hypotheses regarding host/prey partitioning.

288

FUTURE PROSPECTS

To use the house fly system model in actual management situations requires that we include potential management strategies (e.g., pesticide scheduling, manure management alternatives, augmentation of natural enemies, alterations in the temperature and moisture conditions) in terms of their impact on survival and/or emigration for each of the species and their respective stages. The completion of this management model will also necessitate inclusion of economic information, and will be implemented within an expert system frame. The latter is necessary if we are to include subjective information necessary for decisions, but not appropriate for simulation modeling, *per se*. The prospects are promising for the construction of a practical fly management model for use in an expert system. Eventually such a model should be incorporated into a multi-pest management expert system for poultry production as well as be modified for application to other types of confined-animal production systems.

ACKNOWLEDGMENTS

Support for this research has been provided by the North Carolina Agricultural Research Service (NCARS) and Southern Region IPM Grants (Nos. 86-CSRS-2-2889 and 89-341-3-4242) from the U. S. Department of Agriculture, Cooperative State Research Service (USDA-CSRS).

REFERENCES CITED

Ables, J. R. and M. Shepard. 1976. Influence of temperature on oviposition by the parasites *Spalangia endius* and *Muscidifurax raptor*. Environ. Entomol. 5: 511-513.

Ables, J. R., M. Shepard and J. R. Holman. 1976. Development of the parasitoids *Spalangia endius* and *Muscidifurax raptor* in relation to constant and variable temperatures: simulation and validation. Environ. Entomol. 5: 329-332.

Axtell, R. C. 1981. Livestock integrated pest management (IPM): principles and prospects, pp. 31-40. In: F. W. Knapp (ed.), Systems approach to animal health and production. Univ. Kentucky, Lexington.

_____. 1986a. Status and potential of biological control agents in livestock and poultry pest management systems. Entomol. Soc. Am. Misc. Publ. 61: 1-9.

_____. 1986b. Fly management in poultry production: cultural, biological, and chemical. Poultry Sci. 65: 657-667.

_____. 1986c. Fly control in confined livestock and poultry production. Tech. Mongr. Ciba-Geigy Corp., Agric. Div., Greensboro, N.C. 59 pp.

Axtell, R. C. and D. A. Rutz. 1986. Role of parasites and predators as biological fly control agents in poultry production facilities. Entomol. Soc. Am. Misc. Publ. 61: 88-100.

Axtell, R. C. and J. J. Arends. 1990. Ecology and management of arthropods pests of poultry. Annu. Rev. Entomol. 35: 101-126.

Fatchurochim, S., C. J. Geden and R. C. Axtell. 1989. Filth fly oviposition and larval development in poultry manure of various moisture levels. J. Entomol. Sci. 24: 224-231.

Fletcher, M. G., R. C. Axtell and R. E. Stinner. Submitted. Longevity and fecundity of *Musca domestica* (Diptera: Muscidae) as a function of temperature. J. Med. Entomol.

Geden, C. J. and R. C. Axtell. 1988. Predation by *Carcinops pumilio* (Coleoptera: Histeridae) and *Macrocheles muscaedomesticae* (Acarina: Macrochelidae) on the house fly (Diptera: Muscidae): functional response, effects of temperature and availability of alternate prey. Environ. Entomol. 17: 739-744.

Geden, C. J., R. E. Stinner and R. C. Axtell. 1988. Predation by predators of the house fly in poultry manure: effects of predator density, feeding history, interspecific interference, and field conditions. Environ. Entomol. 17: 320-329.

Geden, C. J., R. E. Stinner, D. A. Kramer and R. C. Axtell. 1990. MACMOD: A simulation model for *Macrocheles muscaedomesticae* (Acarina: Macrochelidae) population dynamics and rates of predation on immature house fly (Diptera: Muscidae). Environ. Entomol. 18:in press.

290

Lysyk, T. J. and R. C. Axtell. 1986a. Field evaluation of three methods for monitoring populations of house flies, *Musca domestica* (Diptera: Muscidae), and other filth flies in three types of poultry housing systems. J. Econ. Entomol. 79: 144-151.

_____. 1986b. Movement and distribution of house flies (Diptera: Muscidae) between habitats in two livestock farms. J. Econ. Entomol. 79: 993-998.

_____. 1987. A simulation model of house fly (Diptera: Muscidae) development in poultry manure. Can. Entomol. 119: 427-437.

Mann, J. A., R. E. Stinner and R. C. Axtell. 1990a. Parasitism of house fly, *Musca domestica* L., pupae by four species of Pteromalidae (Hymenoptera): effects of host-parasitoid densities and host distribution. Med. Vet. Entomol. 4: in press.

_____. 1990b. Temperature-dependent development and parasitism rates of four species of Pteromalidae (Hymenoptera) parasitoids of house fly *Musca domestica* L.) pupae. Med. Vet. Entomol. 4: in press.

Mullens, B. A., J. L. Rodriquez and J. A. Meyer. 1987. An epizootiological study of *Entomophthora muscae* in muscoid fly populations in southern California poultry facilities, with emphasis on *Musca domestica*. Hilgardia 55: 1-41.

Patterson, R. S. and D. A. Rutz (eds.). 1986. Biological control of muscoid flies. Entomol. Soc. Am. Misc. Publ. 61: 1-174.

Rueda, L. M. and R. C. Axtell. 1985. Guide to common species of pupal parasites (Hymenoptera: Pteromalidae) of the house fly and other muscoid flies associated with poultry and livestock manure. N.C. Agric. Res. Serv. Tech. Bull. 278: 1-88.

Sharpe, J. H. and D. W. DeMichele. 1977. Reaction kinetics of poikilotherm development. J. Theor. Biol. 64: 649-670.

Stinner, R. E., G. D. Butler, Jr., J. S. Bacheler and C. Tuttle. 1975. Simulation of temperature-dependent development in population dynamics models. Can. Entomol. 107: 1167-1174.

Wagner, T. L., H. I. Wu, P. J. H. Sharpe and R. N. Coulson. 1984. Modeling insect development rates: a literature review and application of a biophysical model. Ann. Entomol. Soc. Am. 77: 208-255.

Weidhaas, D. E. 1986. Models and computer simulations for biological control of flies. Entomol. Soc. Am. Misc. Publ. 61: 57-68.

Weidhaas, D. E., D. G. Haile, P. B. Morgan and G. C. Labrecque. 1977. A model to simulate control of house flies with a pupal parasite, *Spalangia endius*. Environ. Entomol. 6: 489-500.

Wilkerson, G. C., J. W. Mishoe and J. L. Stimac. 1986. Modeling velvetbean caterpillar (Lepidoptera: Noctuidae) populations in soybeans. Environ. Entomol. 15: 809-816.

20. Potential of Biocontrol for Livestock and Poultry Pests

Richard C. Axtell

Livestock and poultry production systems are both biological and managerial systems which have been designed and structured by humans for the perceived benefit of humans. These production systems create artificial man-made ecosystems with consequent problems in arthropod pest management. Management of the arthropod pests is complex and not a simple task. It is complicated by varying pest species, varying animal strains, housing and management practices. It is apparent that arthropod pest management must include an integration of methods in a manner compatible with the animal production practices (Axtell 1981, 1986a). Greater utilization of biological control agents is needed and will occur.

The future of biocontrol in both developed and developing countries depends on at least four major parameters: (1) Social, (2) Economic, (3) Political, and (4) Technical. The last, technical, includes the scientific basis for biocontrol and population management as well as the technology of animal production management. Science and technology will eventually provide the capabilities for better use of biocontrol measures. The other parameters may enhance, limit or negate the use of that science and technology, however. Social constraints, such as attitudes towards animal rights and confined animal production, can result in pressures to change animal production practices in ways which complicate the effective use of biocontrol and necessitate new research. In certain European countries, for example, new regulations on poultry production are causing changes in the size and types of housing which will result in a changed environment for filth fly breeding and the use of biocontrol (Anonymous. 1985). Economic constraints will have a major impact on the development and acceptance of biocontrol measures. Funds from governments and/or industry are subject to many demands and high-tech biotechnology can be a more exciting and enticing investment than biocontrol. The most important economic component, however, is the cost of animal production. Biocontrol must be cost effective for it to be accepted. With increasing costs of insecticides and very limited availability of new insecticides, the opportunities for cost effective use of biocontrol are greater. Political constraints may influence the funding for biocontrol research and the need for and acceptance of biocontrol measures. As a result of the social and political environment, laws and regulations on such things as pesticide registration and use, animal housing, waste disposal, importation of animal products and pesticides, import and export tariffs and taxes are established which may either seriously constrain the use of biocontrol measures or encourage and foster the use of biocontrol. All of these factors make the future for biocontrol uncertain even though promising.

The question is: What is the potential for biocontrol of arthropods affecting livestock and poultry in the developed countries in comparison to the developing countries? To examine this question, let us take an overview and then attempt to focus on some specifics and speculations.

The world food problem is summarized in simplistic terms by recognizing that the developing countries with 70% of the world's population produce only 40% of the food, have only 30% of the world's income, and only 20% of the purchased agricultural inputs. By 1990, 77% of the world's population will probably be located in the developing world. At the same time, it is estimated that 50% of the world's food production, 25% of the world's inputs, and 35% of the world's income will also be in the developing world (Olentine 1982). Thus, there is a vast growth potential in agriculture, notably in livestock and poultry production, in developing countries in the near and far future. Developing countries will not only increase their demand for food products, but will increase their relative ability to produce them. This will include substantial increases in livestock and poultry production with probably the greatest growth in poultry. Animal production in the developing countries will likely become more and more patterned after the production practices now used in the developed countries with particular emphasis on the use of confined, high density facilities. Therefore, the pest problems and management solutions will become more alike in both the developed and developing countries. Consequently, the potential for use of biocontrol in livestock and poultry production in confined animal production facilities is likely to become more similar worldwide. The trends in developed countries will probably apply at a later time to the developing countries.

What is the scope of livestock and poultry production in the world? The data, from many sources, are fragmentary, vary in reliability, and often are only crude estimates. The following data are composites from several sources (Anonymous. 1986, 1987; Olentine 1982; Randell 1985, 1986; Schoeff 1987, 1988; USDA/FAS 1986). In North America (mainly USA) annual amounts of production (in millions, 1982) are: Cattle 174.0, swine 87.3, sheep 22.3, broilers 5.1, and eggs 87.6. In South America (mainly Brazil and Argentina) the amounts are: Cattle 208.6, swine 51.5, sheep 95.2, broilers 1.6, and eggs 17.3. In Western Europe (mainly France, Federal Republic of Germany, Netherlands and UK) the amounts are: Cattle 77.8, swine 78.0, sheep 58.8, broilers 2.4, and eggs 70.5. Annual production amounts in other Western Europe countries (mainly Spain) are: Cattle 14.2, swine 25.9, sheep 19.8, broilers 0.8, and eggs 18.0. With Spain and Portugal more recently joining the European Economic Community (ECC) the region presents a sizeable coordinated block for livestock and poultry production. In Eastern Europe the annual amounts are: Cattle 37.4, swine 70.8, sheep 43.1, broilers 0.8, and eggs 37.6. In the Soviet Union the amounts are: Cattle 115.3, swine 73.5, sheep 141.3, broilers 0.6, and eggs 73.2. Amounts for Asia are poorly documented with lack of hard data for the People's Republic of China and distortions due to the large number of cattle (242.0) recorded for India. Nevertheless, amounts for Asia are: Cattle 270.8, swine 34.8, sheep 90.3, broilers 0.9, and eggs 33.1. For so-called Oceanic Asia (mostly Australia and New Zealand the values are: Cattle 33.5, swine 2.8, sheep 206.0, broilers 0.2, and eggs 3.5. Data were not available for Africa.

Growth in poultry production has been most striking. The growth to meet the needs of the human population is destined to be greatest in China and Asia (Table 1). The developed countries have 26.4% of the human population, but produce 48.2% of the world's poultry and 66.7% of the world's eggs (Randall 1985). Per capita consumption of poultry meat is increasing in the developed countries while consumption of red meats is static or declining (USDA/FAS 1986). For example, in the USA the per capita consumption of poultry in 1987 was 78.6 lbs which was slightly greater (for the first time) than the consumption of beef (Table 2). Economics play a large role. The efficiency of feed conversion of poultry has been steadily increased through selective breeding and improved management so that a feed conversion ratio of 2.0 or less is now

Table 1.   Percent of World Total Numbers of Humans, Poultry and Eggs in One Year in Different Areas.

Area	Humans	Poultry	Eggs
Developed economies	26.4	48.2	66.7
China and others	23.3	12.8	14.4
Asia	29.0	12.8	5.9
Latin America	8.0	14.4	8.1
Sub-Sahara Africa	7.5	6.3	1.9
N. Africa & Middle East	5.8	5.4	3.0

Source: Randall 1985 (1979 data).

Table 2.   Meat Consumption in the United States (lbs. per capita).

Year	Poultry	Beef	Pork
1975	48.6	87.9	50.7
1977	53.2	91.8	55.8
1979	60.5	78.0	63.8
1981	62.4	77.1	65.0
1983	65.1	78.7	62.2
1985	70.1	79.1	62.0
1986	72.5	79.8	58.6
1987	78.6	74.7	59.2

Source: Broiler Industry 1987 (Sept), p. 24.

common in developed countries (Misirlioglu 1987). The marketing of improved breeds through out the world is making poultry production more feasible in developing countries.

These production figures, although rather crude, give an idea of the worldwide distribution of livestock and poultry. Comparisons of figures for several years shows that production is increasing worldwide (Anonymous. 1986, 1987, 1988; Olentine 1982; Randall 1985, 1986; USDA/FAS 1986). There are three trends. First, in recent years the trend has accelerated for the greatest percentage increases to be overall in the developing countries. A second trend is for the increases in both developed and developing countries to be in large operations using high densities of animals and being financed and managed by large companies. A third trend is for the greatest increases in both developed and developing countries to be in poultry production.

What does this mean for the potential of using biocontrol of arthropod pests in livestock and poultry production? Growth in high density, large operations, and especially poultry, results in large concentrations of manure and high numbers of filth flies with consequent increased potential for the use of biocontrol (Naber 1987). The greatest opportunities are, therefore, in facilities, such as cattle feedlots, dairy, swine, and poultry housing, where high densities of animals are produced under a highly structured system of management. The production management systems now in operation in developed countries are very sophisticated and use computerized systems for optimizing feeding and housing conditions. Contract farms under one management (integrator) are forced to adopt the practices dictated by the contractor. This structured management approach is most highly developed in poultry production in the so-called developed countries (Jones 1987), but is rapidly being introduced concurrent with expansion of poultry production in the developing countries. Let us, therefore, use poultry production to examine more closely the potential for using biocontrol. The concepts that apply to poultry production also apply to other high density confined animal production systems, e.g. swine, cattle feed lots and dairy.

Poultry production is integrated (Carter 1985) with one company controlling all phases including the breeder flocks, hatchery, growout houses, feed mills, processing plants and even the packaging, marketing and distribution of the products under one or more trade names. Individual farms operate under contracts which provide specified payments from the parent company to the producer. These payments may be based on numbers of eggs, pounds of meat, feed conversion efficiencies, etc. The parent company actually owns the birds, provides the feed and dictates the management practices on the farm. In this situation, the key to using biocontrol is to convince the parent company of the utility of this approach in their operations. If the parent company is convinced, it will require the contract producers to follow the recommended biocontrol practices. The parent company (integrator) has servicemen who monitor the animal production practices of the contract producers and these servicemen can be trained to monitor the arthropod pests and to institute pest management measures, including the use of biocontrol agents.

The next question is: What pest or pests are most susceptible to the use of biocontrol agents? The greatest progress in developed countries has been in the use of biocontrol agents against filth flies which includes the house fly and related manure-breeding species (Patterson & Rutz 1986). This has even fostered the marketing of fly parasites by some small companies. Successful utilization of fly parasites in an integrated fly management program for poultry, dairy, swine, and feedlots has been difficult and far from being reliable. It is clear that the biocontrol agents must be used as a tool along with cultural measures (manure management) and selective use of insecticides (Axtell 1986a,b). Also, a complex of parasites and predators is needed, not just reliance on a single species. Use of fly pathogens should be considered, but presently research on that topic is far behind investigations of parasites and predators.

At present, biocontrol of filth flies in poultry production (Axtell & Rutz 1986), as well as in cattle and dairy feedlots (Meyer 1986), is being used in some areas of the USA and other developed countries. In the USA it is largely a case of recommending fly management practices that will encourage and enhance the natural populations of fly parasites and predators. Some augmentative releases of parasites are practiced. The basic components of a fly management program for poultry are (Axtell 1986c): (1) Cultural - maintaining the manure as dry as possible in production systems that allow for manure accumulation. Factors in this are minimizing leaking waterers, preventing surface water drainage into the buildings, providing ample air circulation through fans and natural currents unrestricted by outside vegetation. (2) Biological - encouraging the maximum development of indigenous fly parasites and predators by the above manure management practices. A major factor is the recommended practice of only removing the manure partially during cleanout times and always leaving a deep pad of old manure to preserve part of the population of predators and parasites.

The major emphasis is on preserving and enhancing populations of the pteromalid fly parasites, *Muscidifurax* and *Spalangia*, the macrochelid mite predator, *Macrocheles muscaedomesticae*, and the histerid beetle predator, *Carcinops pumulio* (Geden & Axtell 1988; Geden et al. 1988; Rueda & Axtell 1985). (3) Chemical - limited and selective use of insecticides in a manner compatible with preservation of the biocontrol agents is used when necessary. General larviciding of the manure is not recommended since most chemicals kill the beneficial arthropods as readily as the fly larvae. Spot treatment of limited areas of manure that have very high numbers of fly larvae is appropriate. One insecticide, cyromazine, is toxic to fly larvae but relatively harmless to the mite and beetle predators of the flies and, therefore, its use as a feed additive or a manure treatment is compatible with preserving and maintaining the populations of beneficial arthropods in the manure (Axtell & Edwards 1983). Use of chemicals as baits or limited residual surface applications directed against the adult flies is appropriate.

These components of an integrated fly management program require a fly population monitoring program. Favored monitoring tools are spot cards and baited jug traps (Lysyk & Axtell 1985, 1986). Weekly counts from these monitoring tools are used by the poultry servicemen and producer to decide on fly control measures. With this attention to actual fly numbers, it has become possible to limit the amount of insecticide used and hence the cost of fly control. Hopefully, this also will slow down the rate of insecticide resistance development by the flies. Practical application of such an integrated fly management program has necessitated the training of servicemen for the major poultry integrators. Similar programs are in use for fly management around cattle feedlots and dairies in the USA and elsewhere.

To more effectively use these biocontrol agents against filth flies, the complex system requires development of some computer models to aid in decision-making by the integrator and producer. Using the poultry production system as an example, we are attempting to model the fly management system. The dynamics of each of the components is complicated. One needs to look at six components: (1) House fly (the major target pest), (2) other fly species, (3) pteromalid fly parasites (population dynamics of *Muscidifurax* and *Spalangia*), (4) predators (*Macrocheles* mites and *Carcinops* beetles), (5) fly pathogens, and (6) the nature of the environment. The last component is critical for it involves the temperature (which is the major factor driving the arthropod population submodels) and manure quality (which affects fly breeding and populations of biocontrol agents). Temperature and manure quality are affected by the type of animal housing, types of animals, animal physiology and nutrition, animal management practices and the manure handling and disposal practices.

Putting all of these factors together in computer simulation models for fly management in poultry or other confined animal production facilities will be a formidable task, but progress is being made in developed countries and this technology will be transferred with the introduction of highly integrated animal production technology into other countries. It is likely that the computer simulation models will not be developed to handle all aspects of the fly management problem. The data are not available and are too difficult to obtain for many aspects of the problem. The solution to this will be to develop so-called "expert systems" which will use the judgement of persons knowledgeable and experienced in the field of fly control along with inputs from the computer simulation models that are successfully developed. These expert systems will provide a structured program for decision-making in regards to fly management that will be compatible with the computer tools already being used by integrators for managing their poultry and livestock production enterprises. Use of computers has proven beneficial and cost effective to the integrators and, consequently, will be readily accepted and used if the fly management expert systems really work. With such systems, it should be possible to use biocontrol agents effectively in conjunction with manure management and insecticides.

The key to using biocontrol agents for filth fly control is manure management. The amount of manure produced per animal is inescapable and unless handled properly will be a source of uncontrollable fly breeding. The pounds of manure per day per 100 lbs of animal weight are: Dairy cow 8.2, beef cattle 6.0, swine 6.5, sheep 4.0, horse 4.5, laying hen 5.3, and broiler hen 7.1. Expressed in another way, poultry excrete about 5% of their body weight per day and the manure contains 75% moisture. Cattle (dairy and beef) and swine excrete 7-8% of their body weight per day and the manure contains 80-85% moisture. The handling and disposal of such large quantities of manure from confined livestock and poultry is a serious problem. The use of manure as fertilizer is a subject of continuing development. If the practical problems of drying, handling and distributing can be solved, then manure will become a positive benefit of increased livestock and poultry production (Anonymous. 1985, Naber 1987). The need for fertilizers is increasing while animal production is increasing. To illustrate the point, the growth in fertilizer use from 1975 to 1982 worldwide has occurred while there

Table 3. Egg Production and Nitrogen Fertilizer Use (Percent of World Totals) in Different Regions.

Region	Eggs 1975	Eggs 1982	Fertilizer 1975	Fertilizer 1982
North America	20.1	18.1	25.5	21.1
United States	15.9	13.7	21.7	16.6
Europe	27.3	24.5	28.1	23.9
USSR	13.2	13.3	16.9	13.9
Asia	30.8	33.8	24.5	35.6
South America	4.5	6.0	5.6	1.9
Africa	3.0	3.4	2.8	3.1
TOTAL (million metric tons)	24.04	29.84	43.56	60.42

Source: World Poultry 1985 (Dec), p. 62.

has been a similar growth in egg production (Table 3). Manure may be used for methane production and as a component of animal feed, but these uses have not become widely adopted. In developed countries, the lack of sufficient crop land for manure disposal is limiting the expansion of poultry production as well as other confined animal operations. Disposal of manure in deep lagoons is a temporary expedient, but eventually the decomposed organic matter has to be removed from the lagoons and in the meantime the lagoons frequentlycreate a mosquito problem due to the ideal conditions for breeding *Culex quinquefasciatus* (Rutz & Axtell 1978; Rutz et al. 1980). If the lagoon disposal method is introduced into tropical Asian countries, the consequent populations of *C. quinquefasciatus* could be disastrous because that species is the major vector of human filariasis (*Wuchereria bancrofti*). The potential use of biocontrol agents for filth fly control is strongly linked to the manure management practices being developed and adopted in developed countries.

The potential of biocontrol for arthropods affecting livestock and poultry in developed countries is great in the case of integration into filth fly management programs for confined animal production facilities. The merits of using parasites and predators in such programs has been demonstrated empirically and quantitative data are being developed. The concept is being accepted and put into prac-

302

tice by the segments of the poultry industry as well as in cattle feedlots and dairy operations. The use of biocontrol agents for other arthropods and other categories of animal production is much less advanced and appears less promising. Future application of biocontrol will depend upon expansion and improvement of our scientific and technical knowledge and whether or not the social, economic and political conditions are favorable.

REFERENCES CITED

Anonymous. 1985a. Welfare symposium: Costs of production are the only clear guide for choice of laying systems. World Poultry (Aug), pp. 34-35.

Anonymous. 1985b. Manure drying: Swedish drier manufacturer claims a sterile product. World Poultry (Dec), p. 62.

Anonymous. 1986. ECC facts and forecasts: Squeeze on the community. World Poultry (Oct), pp. 53, 54, 59.

Anonymous. 1987. EEC facts and forecasts. World Poultry (Oct), pp. 19-24.

Anonymous. 1988. World data: More poultry feeds world appetites. World Poultry (Feb), pp. 28, 30.

Axtell, R. C. 1981. Livestock integrated pest management (IPM): Principles and prospects. p. 34-40 In: Systems Approach to Animal Health and Production: A Symposium (F. W. Knapp, ed.), Univ. Kentucky, Lexington. 241 pp.

_____. 1986a. Status and potential of biological control agents in livestock and poultry pest management systems. pp. 1-9 In: Biological Control of Muscoid Flies (Patterson, R.S. & D. A. Rutz, eds.) Misc Publ. Entomol. Soc. Amer. 61, 174 pp.

_____. 1986b. Fly control in confined livestock and poultry production. Technical Monograph, Ciba-Geigy Corp., Greensboro, NC. 59 pp.

_____. 1986c. Fly management in poultry production: Cultural, biological and chemical. Poultry Science 65: 657-667.

Axtell, R. C. & T. D. Edwards. 1983. Efficacy and non-target effects of Larvadex as a feed additive for controlling house flies in caged-layer poultry manure. Poultry Science 62: 2371-2377.

Axtell, R. C. & D. A. Rutz. 1986. Role of parasites and predators as biological fly control agents in poultry production facilities. p. 88-100 In: Biological Control of Muscoid Flies (Patterson, R. S. & D. A. Rutz, eds.) Misc. Publ. Entomol. Soc. Amer. 61, 174 pp.

Carter, T. A. 1985. The North Carolina poultry industry. Poultry Sci. & Tech. Guide No. 39, North Carolina Agric. Ext. Serv., 6 pp.

Geden, C. J. & R. C. Axtell. 1988. Predation by *Carcinops pumilio* (Coleoptera: Histeridae) and *Macrocheles muscaedomesticae* (Acarina: Macrochelidae) on the house fly (Diptera: Muscidae): Functional response, effects of temperature, and availability of alternative prey. Environ. Entomol. 17: 739-744.

Geden, C. J., R. E. Stinner & R. C. Axtell. 1988. Predation by predators of the house fly in poultry manure: Effects of predator density, feeding history, interspecific interference, and field conditions. Environ. Entomol. 17: 320-329.

Jones, H. B. 1987. Broiler stocks: Is the uptrend over? Broiler Industry (Sept), pp. 42, 44, 46, 50.

Lysyk, T. J. & R. C. Axtell. 1985. Comparison of baited jug-trap and spot cards for sampling house fly, *Musca domestica* (Diptera: Muscidae), populations in poultry houses. Environ. Entomol. 14: 815-819.

_____. 1986. Field evaluation of three methods for monitoring populations of house flies (*Musca domestica*) (Diptera: Muscidae) and other filth flies in three types of poultry housing systems. J. Econ. Entomol. 79: 144-151.

Meyer, J. A. 1986. Biological control of filth flies associated with confined livestock. p. 108-115 In: Biological Control of Muscoid Flies (Patterson, R. S. & D. A. Rutz, eds.) Misc. Publ. Entomol. Soc. Amer. 61, 174 pp.

Misirlioglu, H. S. 1987. Economics of feed conversion. Broiler Industry (Sept), pp. 76-81.

Naber, E. C. 1987. Poultry manure utilization and management task force summary. Zootecnica International (Jan), pp. 32, 34, 36.

Olentine, C. 1982. World feed prospects. Feed Management (July), pp. 22, 24, 26, 28.

Patterson, R. S. & D. A. Rutz (eds.). 1986. Biological control of muscoid flies. Misc. Publ. Entomol. Soc. Amer. 61, 174 pp.

Randall, K. 1985. Food protein: Third world production trends indicate a meat deficit but enough eggs by 2000. World Poultry (Sept), p. 30.

_____. 1986. South east Asia: Increased output fills gains in poultry and egg consumption. World Poultry (Aug), p. 9.

Rueda, L. M. & R. C. Axtell. 1985. Guide to common species of pupal parasites (Hymenoptera: Pteromalidae) of the house fly and other muscoid flies associated with poultry and livestock manure. North Carolina Agric. Research Serv. Bull. 278, 88 pp.

Rutz, D. A. & R. C. Axtell. 1978. Factors affecting production of the mosquito, *Culex quinquefasciatus* (= *fatigans*) from anaerobic animal waste lagoons. North Carolina Agric. Expt. Sta. Tech. Bull. 256, 32 pp.

Rutz, D. A., R. C. Axtell & T. D. Edwards. 1980. Effects of organic pollution levels on aquatic insect abundance in field pilot-scale anaerobic animal waste lagoons. Mosquito News 40: 403-409.

Schoeff, R. 1987. Market data 1987. Feed Management (Oct), pp. 20, 22, 24, 26, 28.

_____. 1988. Market data 1988. Feed Management (Oct), pp. 40, 42-48.

USDA/FAS. 1986 (March). Dairy, livestock and poultry: World livestock and poultry situation. Foreign Agric. Circular, USDA, Foreign Agric. Service, FL&P1-86.

Contributors

Dr. Richard S. Patterson, United States Department of Agriculture, Agricultural Research Service, Box 14565, Gainesville, FL 32604

Dr. M. Geetha Bai, Karnataka State Sericulture Developmental Institute, Thalaghattapura, Bangalore 560062 INDIA

Dr. Gerald L. Greene, Southwest Kansas Branch Experiment Station, 4500 E. Mary, Garden City, KS 67846-9132

Dr. Jeffery A. Meyer, College of Natural & Agricultural Sciences, Department of Entomology, University of California, Riverside, CA 92521

Dr. Richard W. Miller, Agricultural Research Service/United States Department of Agriculture, Bldg. 177a, BARC-E, Beltsville, MD 20705

Dr. Philip B. Morgan, United States Department of Agriculture, Agricultural Research Service, Box 14565, Gainesville, FL 32604

Dr. James J. Petersen, United States Department of Agriculture, Agricultural Research Service, Midwest LIL, University of Nebraska, Lincoln NE 68506

Dr. D. W. Watson, United States Department of Agriculture, Agricultural Research Service, Midwest LIL, University of Nebraska, Lincoln NE 68506

Dr. B. M. Pawson, United States Department of Agriculture, Agricultural Research Service, Midwest LIL, University of Nebraska, Lincoln NE 68506

Dr. Martin J. Rice, University of Queensland, St. Lucia Queensland, AUSTRALIA 4067

Dr. Renato Ripa S., Subestacion Experimental La Cruz, Agriculture Research Institute, La Cruz, CHILE

Dr. Leopoldo M. Rueda, North Carolina State University, Box 7613, Department of Entomology, Raleigh, NC 27695-7613

Dr. C. T. Hugo, North Carolina State University, Box 7613, Department of Entomology, Raleigh, NC 27695-7613

Dr. M. B. Zipagan, North Carolina State University, Box 7613, Department of Entomology, Raleigh, NC 27695-7613

Dr. G. Truman Fincher, United States Department of Agriculture, Agricultural Research Service, Veterinary Toxicology and Entomology Research Laboratory, P.O. Drawer GE, College Station, TX 77841

Dr. E.T.E. Darwish, Plant Protection Department, Menoufia University, Shebin El-Kom EGYPT

Dr. A. M. Zaki, Plant Protection Department, Menoufia University, Shebin El-Kom EGYPT

Dr. A. A. Osman, Plant Protection Department, Menoufia University, Shebin El-Kom EGYPT

Dr. Róbert Farkas, Department. of General Zoology & Parasitology, University Veterinary Science, Landler J.U.2, 1078 Budapest, HUNGARY

Dr. László Papp, Department. of General Zoology & Parasitology, University Veterinary Science, Landler J.U.2, 1078 Budapest, HUNGARY

Dr. Christopher J. Geden, Department of Entomology, Comstock Hall, Cornell University, Ithaca, NY 14853-0999

Dr. Chyi-Chen Ho, Taiwan Agricultural Research Institute, Wufeng Taichung, Taiwan, REPUBLIC OF CHINA

Dr. Harvey L. Cromroy, Taiwan Agricultural Research Institute, Wufeng Taichung, Taiwan, REPUBLIC OF CHINA

Dr. Jorgen B. Jespersen, Danish Pesticide Information Laboratory, Skovbrynet 14, DK-2800, Lyngby, DENMARK

Dr. J. Keiding, Danish Pesticide Information Laboratory, Skovbrynet 14, DK-2800, Lyngby, DENMARK

Dr. Bradley A. Mullens, Department of Entomology, University of California, Riverside, CA 92521

Dr. Donald A. Rutz, Department of Entomology, Comstock Hall, Cornell University, Ithaca, NY 14853-0999

Dr. Jeffrey G. Scott, Department of Entomology, Comstock Hall, Cornell University, Ithaca, NY 14853-0999

Dr. Richard C. Axtell, North Carolina State University, Box 7613, Department of Entomology, Raleigh, NC 27695-7613

Dr. R. E. Stinner, North Carolina State University, Box 7613, Department of Entomology, Raleigh, NC 27695-7613

Index

Printed in the United States
by Baker & Taylor Publisher Services